Plane Finite Elements for Two-Dimensional Problems

Andreas Öchsner · Resam Makvandi

Plane Finite Elements for Two-Dimensional Problems

Application of the Computer Algebra System Maxima

Springer

Andreas Öchsner
Faculty of Mechanical Engineering
Esslingen University of Applied Sciences
Esslingen am Neckar
Baden-Württemberg, Germany

Resam Makvandi
Institute of Mechanics
Otto von Guericke University Magdeburg
Magdeburg, Sachsen-Anhalt, Germany

ISBN 978-3-030-89549-5 ISBN 978-3-030-89550-1 (eBook)
https://doi.org/10.1007/978-3-030-89550-1

This Springer imprint is published by the registered company Springer Nature Switzerland AG
The registered company address is: Gewerbestrasse 11, 6330 Cham, Switzerland

Only a generation of readers will spawn a generation of writers.

—Steven Spielberg

Preface

This book is intended as a study aid for the finite element method. Based on the free computer algebra system Maxima, we offer routines to symbolically or numerically solve problems from the context of two-dimensional problems. For this rather advanced topic, classical 'hand calculations' are difficult to perform and the incorporation of a computer algebra system is a convenient approach to handle, for example, larger matrix operations. The mechanical theories focus on the classical two-dimensional structural elements, i.e., plane elements, thin or classical plates, and thick or shear deformable plate elements. The use of a computer algebra system and the incorporated functions, e.g., for matrix operations, allows to focus more on the methodology of the finite element method and not on standard procedures. Furthermore, we offer a graphical user interface (GUI) to facilitate the model definition. Thus, the user may enter the required definitions in a source code manner directly in wxMaxima or use the GUI which is able to execute wxMaxime to perform the calculations.

We look forward to receiving some comments and suggestions for the next edition of this textbook.

Esslingen am Neckar, Germany — Andreas Öchsner
Magdeburg, Germany — Resam Makvandi
August 2021

Acknowledgement We would like to express our sincere appreciation to the Springer Publisher, especially to Dr. Christoph Baumann, for giving us the opportunity to realize this book.

Contents

Symbols and Abbreviations

Latin Symbols (Capital Letters)

A Area, cross-sectional area
$\boldsymbol{B}$ Matrix which contains derivatives of interpolation functions
$\boldsymbol{C}$ Elasticity matrix
D Bending rigidity
$\boldsymbol{D}$ Plate elasticity matrix
$\boldsymbol{D}^*$ Plate elasticity matrix
E Young's modulus
F Force
G Shear modulus
J Jacobian determinant
K Stiffness matrix element
$\boldsymbol{K}^{\mathrm{e}}$ Elemental stiffness matrix
L Length
M Moment
N Interpolation function
$\boldsymbol{N}$ Column matrix of interpolation functions
$\overline{N}$ Shape function
P Reference point, Degree of polynomial
Q Shear force (internal)
S Surface
V Volume
X Global Cartesian coordinate
Y Global Cartesian coordinate
Z Global Cartesian coordinate

Latin Symbols (Small Letters)

a	Geometric dimension
b	Geometric dimension
d	Geometric dimension
e	Column matrix of generalized strains
f	Function, body force
$\boldsymbol{f}$	Global column matrix of nodal loads
$\boldsymbol{f}^{\mathrm{e}}$	Elemental column matrix of nodal loads
h	Plate thickness
i	Node number
k_{s}	Shear correction factor
m	Distributed moment
n	Number of quadrature point
p	Distributed load
q	Distributed load
s	Local coordinate
s	Column matrix of generalized stresses
t	Plane thickness, traction force
u	Displacement
$\boldsymbol{u}$	Global column matrix of nodal deformations
$\boldsymbol{u}^{\mathrm{e}}$	Elemental column matrix of nodal deformations
w	Weight for numerical integration
x	Cartesian coordinate
y	Cartesian coordinate
z	Cartesian coordinate

Greek Symbols (Small Letters)

α	Numerical factor
γ	Shear strain
ε	Normal strain
ε	Column matrix of strain components
ζ	Natural coordinate
η	Natural coordinate
ν	Poisson's ratio
ξ	Natural coordinate
σ	Normal stress
$\boldsymbol{\sigma}$	Column matrix of stress components
ϕ	Rotation
φ	Rotation

Mathematical Symbols

∂	Partial derivative symbol (rounded *d*)
$\times$	Multiplication sign (used where essential)
$\mathcal{L}\{\dots\}$	Differential operator
$\boldsymbol{\mathcal{L}}$	Matrix of differential operators

Indices, Superscripted

$\dots^{\mathrm{e}}$	Element
$\dots^{\mathrm{n}}$	Length-specific
$\dots^{\mathrm{T}}$	Transpose

Indices, Subscripted

$\dots_{\mathrm{b}}$	Bending
$\dots_{\mathrm{p}}$	Point
$\dots_{\mathrm{s}}$	Shear

Abbreviations

API	Application programming interface
CAD	Computer-aided design
FE	Finite element
FEM	Finite element method
GNU	GNU's Not Unix
GPL	General public license
GUI	Graphical user interface

Chapter 1
Introduction

The finite element method (FEM) can be considered as the most important computer tool in structural mechanics. Despite the fact that there are other numerical approximation techniques known in literature, e.g., the finite difference method or the boundary element method, the FEM is the most common tool in industrial practice. The method itself has practically no limitations in regard to the physical problem. However, the available computer hardware, i.e., working memory and processing power, may impose some limitations on the size of the problems. A diversity of commercial finite element packages with different specializations is nowadays available for the calculation engineer. Some finite element packages are even incorporated in computer-aided design (CAD) programs. This facilitates the conversion of a geometry into a finite element mesh. However, a reliable application of this powerful tool requires an in-depth understanding of the underlying theory. The challenge is not to obtain from a commercial package some numbers or colorful pictures. The challenge and task of an engineer is to guarantee good results.

To study the finite element method is a challenging task and different approaches are available. This ranges from classical lectures and the corresponding textbooks [1, 2, 6, 12, 13] to the classical tutorials with 'hand calculations' [5]. In a more modern academic context, the so-called problem or project based approaches are also common in some countries [7]. It is also common that the more theoretical sessions are accompanied by computer based laboratories where a commercial finite element package is introduced [10]. As an alternative, a real programming language can be used to teach the computer implementation and to develop own routines [4, 11]. The scope of this book is slightly different since we are going to use the functionality of a free computer algebra system. Doing so, we can avoid the need to program standard routines and focus more on the methodology. The idea for this book partly stems from lecture notes which are based on the commercial computer algebra system Mathematica [3]. However, we decided to use the free computer algebra system Maxima to avoid any license issues since the software can be freely downloaded for all common operating systems. Furthermore, different graphical user interfaces can be used. The successful application of Maxima for finite element applications has already been shown by the authors of this book for one-dimensional problems [8, 9].

A. Öchsner and R. Makvandi, *Plane Finite Elements for Two-Dimensional Problems*,
https://doi.org/10.1007/978-3-030-89550-1_1

References

1. Bathe K-J (1996) Finite element procedures. Prentice-Hall, Upper Saddle River
2. Cook RD, Malkus DS, Plesha ME, Witt RJ (2002) Concepts and applications of finite element analysis. Wiley, New York
3. Introduction to finite element methods (ASEN 5007) Fall 2017. Department of Aerospace Engineering Sciences, University of Colorado at Boulder. https://www.colorado.edu/engineering/CAS/courses.d/IFEM.d/. Accessed 25 Mar 2018
4. Javanbakht Z, Öchsner A (2017) Advanced finite element simulation with MSC Marc: application of user subroutines. Springer, Cham
5. Javanbakht Z, Öchsner A (2018) Computational statics revision course. Springer, Cham
6. Öchsner A (2020) Computational statics and dynamics: an introduction based on the finite element method. Springer, Singapore
7. Öchsner A (2021) A project-based introduction to computational statics. Springer, Cham
8. Öchsner A, Makvandi R (2019) Finite elements for truss and frame structures: an introduction based on the computer algebra system Maxima. Springer, Cham
9. Öchsner A, Makvandi R (2019) Finite elements using Maxima: theory and routines for rods and beams. Springer, Cham
10. Öchsner A, Öchsner M (2018) A first introduction to the finite element analysis program MSC Marc/Mentat. Springer, Cham
11. Trapp M, Öchsner A (2018) Computational plasticity for finite elements: a Fortran-based introduction. Springer, Cham
12. Zienkiewicz OC, Taylor RL (2000) The finite element method. Vol. 1: the basis. Butterworth-Heinemann, Oxford
13. Zienkiewicz OC, Taylor RL (2000) The finite element method. Vol. 2: solid mechanics. Butterworth-Heinemann, Oxford

Chapter 2
Maxima—A Computer Algebra System

The computer algebra system Maxima was originally developed in the late 1960s and earlier 1970s at MIT under the name 'Macsyma'. The historical development is excellently summarized in the article by Moses [2]. Compared to many commercial alternatives such as Maple, Matlab or Mathematica, Maxima is distributed under the GNU General Public License (GPL) and thus is a free software package. In addition, it is recommended to use in the following the wxMaxima graphical user interface (GUI), which is also distributed under the GNU GPL. The routines used in the following chapters are optimized for this GUI. Both programs can be downloaded from the links in Table 2.1.

A few basic operations in Maxima are explained in the following without any attempt for completeness. The interested reader may find elsewhere further links and examples in order to study the functionality of Maxima [1]. Particularly worth mentioning is the web page 'Maxima by Example' by Woollett [3]. Let us now start with the basic arithmetic operations as outlined in the following listing.

Table 2.1 Links to download the installation packages

Program	Link
Maxima	https://maxima.sourceforge.io/download.html
wxMaxima	https://wxmaxima-developers.github.io/wxmaxima/

A. Öchsner and R. Makvandi, *Plane Finite Elements for Two-Dimensional Problems*,
https://doi.org/10.1007/978-3-030-89550-1_2

Table 2.2 Some pre-defined functions in Maxima

Command	Meaning	Command	Meaning
sqrt(...)	square root	sin(...)	sine
exp(...)	exponential	cos(...)	cosine
log(...)	natural logarithm	tan(...)	tangent
abs(...)	absolute value	cot(...)	cotangent

Basic arithmetics: +, −, ∗, /

```
(%i1)  1 + 2;
(%o1)  3

(%i2)  1 - 2;
(%o2)  -1

(%i3)  1 * 2;
(%o3)  2

(%i4)  1 / 2;
(%o4)  1/2
```

We can see from the above example that all expressions entered into Maxima must end with a semicolon ';'. Alternatively, the '$' character can be used at the end of a statement to suppress the output of that line. Some pre-defined functions in Maxima are collected in Table 2.2. It should be noted here that the arguments of the trigonometric functions must be given in radians.

Another characteristic is that Maxima tries to provide symbolic results, which include fractions, square roots etc. The function float(...) can be used to obtain a floating-point representation as shown in the following listing. The most recent result can be recalled with the percentage operator (%).

```
(%i1)   1/2;
(%o1)   1/2

(%i2)   float(%);
(%o2)   0.5

(%i3)   float(1/2);
(%o3)   0.5
```

The value of a variable is entered by the use of a colon ':', see the following listing. To clear a value from an assigned variable, the command kill(...), or for all variables kill(all), is used.

```
Definition of variables and calculations:
(%i2)   a: 3;
        b: 4;

(a)     3
(b)     4

(%i3)   c: a + b;

(c)     7
```

Some predefined constants in Maxima are '%e' (i.e., the base of the natural logarithm; e = 2.718281...), '%pi' (i.e., the ratio of the perimeter of a circle to its diameter; $\pi = 3.141592...$), and '%i' (i.e., the imaginary unit; $\sqrt{-1}$).

A function is defined by the use of a colon followed by the equal sign ':=', see the following listing. In regard to the naming of variables and functions, it should be noted that the names must start with a letter and may contain numbers or even an underscore. The following way of defining a function uses the general structure

f(x) := (expr1, expr2, ..., exprn),

where the value of exprn is returned by the function $f(x)$.

```
Purpose: Calculates the sum of two numbers.
  Input(s): Values a and b.
  Output: Sum of a and b stored in variable c

(% i1)   kill(all)$
(% i1)   summation(a,b) := a+b $
(% i3)   a:3 $
         b:5 $
(% i4)   c : summation(a,b);

(c)      8
```

The following example shows the definition of a function under consideration of a block structure. The block structure allows to make a return from some expression inside the function, to return multiple values, and to introduce local variables.

```
Purpose: Calculates the length of a straight line between two points.
  Input(s): Coordinates of the end points.
  Output(s): Length of the line connecting the points.

-->      LineLength(ncoor):=
         block([x1,x2,y1,y2,x21,y21,L,LL],
         [[x1,y1],[x2,y2]] : ncoor,
         [x21,y21] : [x2-x1,y2-y1],
         LL : (x21^2+y21^2),
         L : sqrt(LL),
         return(L)
         )$

-->      LineLength([[0,0],[1,1]]);

(% o2)   √2
```

Let us mention at the end of this section the command ratsimp(...), which allows to simplify algebraic expressions.[1] The following listing illustrates the simplification of the expressions

$$f_1(x) = 2x^2 + (x-1)^2\,, \tag{2.1}$$

$$f_2(x) = \frac{x^2 + 4x - 3x^2 + 1}{(x-1)^2 - 1 + 2x}\,. \tag{2.2}$$

[1] Further commands for the manipulation of algebraic expressions can be found in the references given on the web page [1].

Manipulation of algebraic expressions:

(% i2)
```
f1: 2*x^2+(x-1)^2$
f2: (x^2+4*x-3*x^2+1)/((x-1)^2-1+2*x)$
```

(% i3)
```
ratsimp(f1);
```

(% o3) $3x^2 - 2x + 1$

(% i4)
```
ratsimp(f2);
```

(% o4) $-\frac{2x^2-4x-1}{x^2}$

References

1. Maxima Dokumentation. https://maxima.sourceforge.io/documentation.html. Cited 23 March 2018
2. Moses J (2012) Macsyma: a personal history. J Symb Comput 47:123–130
3. Woollett EL (2018) Maxima by example. http://web.csulb.edu/~woollett/. Cited 23 March 2018

Chapter 3
Plane Elements

3.1 Calculation of the Elemental Stiffness Matrix

A plane elasticity element is defined as a thin two-dimensional member, as schematically shown in Fig. 3.1, with a much smaller thickness t than the planar dimensions. It can be seen as a two-dimensional extension or generalization of a one-dimensional rod element. The following derivations are restricted to some simplifications:

- the thickness t is constant or a function $t = t(x, y)$ but much smaller than the planer dimensions a and b,
- the undeformed member shape is planar,
- the material is isotropic, homogeneous and linear-elastic according to HOOKE's law for a plane stress or plane strain state,
- external forces act only at the boundary parallel to the plane of the member,
- stresses from external forces are assumed to be distributed uniformly over the thickness.

In regard to the constitutive equation for plane elasticity problems, we are going to distinguish the plane stress and the plane strain state.

The *plane stress* case ($\sigma_z = \sigma_{yz} = \sigma_{xz} = 0$) is commonly used for the analysis of thin, flat plates loaded in the plane of the plate (x-y plane). It is important to emphasize that the normal thickness stress is zero ($\sigma_z = 0$) whereas the thickness normal strain is present ($\varepsilon_z \neq 0$). The plane stress Hooke's law for a linear-elastic isotropic material based on the Young's modulus E and Poisson's ratio ν can be written for a constant temperature as

$$\begin{bmatrix} \sigma_x \\ \sigma_y \\ \sigma_{xy} \end{bmatrix} = \frac{E}{1-\nu^2} \begin{bmatrix} 1 & \nu & 0 \\ \nu & 1 & 0 \\ 0 & 0 & \frac{1-\nu}{2} \end{bmatrix} \begin{bmatrix} \varepsilon_x \\ \varepsilon_y \\ 2\,\varepsilon_{xy} \end{bmatrix}, \tag{3.1}$$

or in matrix notation as

$$\boldsymbol{\sigma} = \boldsymbol{C}\boldsymbol{\varepsilon}\,, \tag{3.2}$$

A. Öchsner and R. Makvandi, *Plane Finite Elements for Two-Dimensional Problems*,
https://doi.org/10.1007/978-3-030-89550-1_3

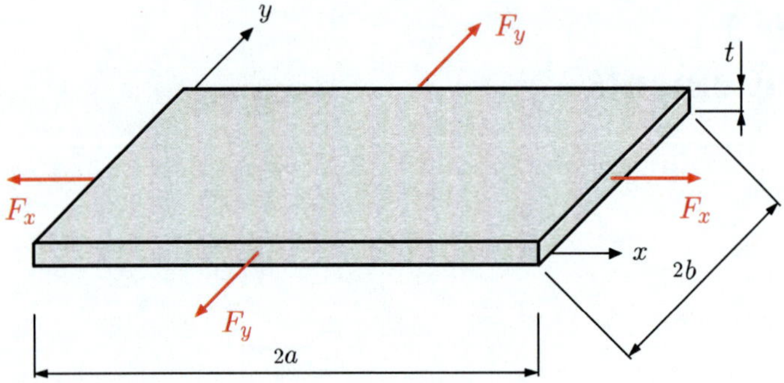

Fig. 3.1 General configuration for a plane elasticity problem

where $\boldsymbol{C}$ is the so-called elasticity matrix. It should be noted here that the engineering shear strain $\gamma_{xy} = 2\varepsilon_{xy}$ is used in the formulation of Eq. (3.1). It should be finally noted that the thickness strain ε_z can be obtained based on the two in-plane normal strains ε_x and ε_y as:

$$\varepsilon_z = -\frac{\nu}{1-\nu} \cdot \left(\varepsilon_x + \varepsilon_y\right). \tag{3.3}$$

The last equation can be derived from the general three-dimensional formulation, see [4].

The *plane strain* case ($\varepsilon_z = \varepsilon_{yz} = \varepsilon_{xz} = 0$) is commonly used for the analysis of elongated prismatic bodies of uniform cross-section subjected to uniform loading along their longitudinal axis but without any component in direction of the z-axis (e.g., pressure), such as in the case of tunnels, soil slopes, and retaining walls. It is important to emphasize that the normal strain in the thickness direction is zero ($\varepsilon_z = 0$) whereas the normal stress in the thickness direction is present ($\sigma_z \neq 0$). The plane strain Hooke's law for a linear-elastic isotropic material based on the Young's modulus E and Poisson's ratio ν can be written for a constant temperature as

$$\begin{bmatrix} \sigma_x \\ \sigma_y \\ \sigma_{xy} \end{bmatrix} = \frac{E}{(1+\nu)(1-2\nu)} \begin{bmatrix} 1-\nu & \nu & 0 \\ \nu & 1-\nu & 0 \\ 0 & 0 & \frac{1-2\nu}{2} \end{bmatrix} \cdot \begin{bmatrix} \varepsilon_x \\ \varepsilon_y \\ 2\,\varepsilon_{xy} \end{bmatrix}, \tag{3.4}$$

or in matrix notation as

$$\boldsymbol{\sigma} = \boldsymbol{C}\boldsymbol{\varepsilon}\,, \tag{3.5}$$

where $\boldsymbol{C}$ is again the so-called elasticity matrix. It should be finally noted that the thickness stress σ_z can be obtained based on the two in-plane normal stresses σ_x and σ_y as:

$$\sigma_z = \nu(\sigma_x + \sigma_y)\,. \tag{3.6}$$

The last equation can be derived from the tree-dimensional formulation, see [4].

3.1.1 Elements with Four Nodes

Let us consider in the following the four-node planar element as shown in Fig. 3.2. Important for the following derivations is that the node numbering is counterclockwise. Furthermore, all the following derivations are restricted to a linear four-node element under the plane stress or plane strain assumption.

The general expression for the elemental stiffness matrix $\boldsymbol{K}^{\mathrm{e}}$ of a four-node plane elasticity element with elasticity matrix $\boldsymbol{C}$ can be expressed as

$$\boldsymbol{K}^{\mathrm{e}} = \int_V \underbrace{(\mathcal{L}_1\boldsymbol{N}^{\mathrm{T}})^{\mathrm{T}}}_{\boldsymbol{B}} \boldsymbol{C} \underbrace{(\mathcal{L}_1\boldsymbol{N}^{\mathrm{T}})}_{\boldsymbol{B}^{\mathrm{T}}} \mathrm{d}V\,, \tag{3.7}$$

where the $\boldsymbol{B}$-matrix contains the derivatives of the interpolation functions N_i:

$$\boldsymbol{B} = (\mathcal{L}_1\boldsymbol{N}^{\mathrm{T}})^{\mathrm{T}} = \left(\begin{bmatrix} \frac{\partial}{\partial x} & 0 \\ 0 & \frac{\partial}{\partial y} \\ \frac{\partial}{\partial y} & \frac{\partial}{\partial x} \end{bmatrix} \left[\begin{array}{cc|cc|cc|cc} N_1 & 0 & N_2 & 0 & N_3 & 0 & N_4 & 0 \\ 0 & N_1 & 0 & N_2 & 0 & N_3 & 0 & N_4 \end{array}\right]\right)^{\mathrm{T}} \tag{3.8}$$

$$= \left(\left[\begin{array}{cc|cc|cc|cc} \frac{\partial N_1}{\partial x} & 0 & \frac{\partial N_2}{\partial x} & 0 & \frac{\partial N_3}{\partial x} & 0 & \frac{\partial N_4}{\partial x} & 0 \\ 0 & \frac{\partial N_1}{\partial y} & 0 & \frac{\partial N_2}{\partial y} & 0 & \frac{\partial N_3}{\partial y} & 0 & \frac{\partial N_4}{\partial y} \\ \frac{\partial N_1}{\partial y} & \frac{\partial N_1}{\partial x} & \frac{\partial N_2}{\partial y} & \frac{\partial N_2}{\partial x} & \frac{\partial N_3}{\partial y} & \frac{\partial N_3}{\partial x} & \frac{\partial N_4}{\partial y} & \frac{\partial N_4}{\partial x} \end{array}\right]\right)^{\mathrm{T}}, \tag{3.9}$$

which is a (8×3)-matrix. The transposed, i.e., $\boldsymbol{B}^{\mathrm{T}} = \mathcal{L}_1\boldsymbol{N}^{\mathrm{T}}$, is thus a (3×8)-matrix. The triple matrix multiplication gives after integration the stiffness matrix with a dimension of 8×8.

The integration for the element matrix given in Eq. (3.7) is approximated by numerical integration in commercial finite element packages. To this end, the coordinates (x, y) are transformed to the natural coordinates (unit space: ξ, η) where each coordinate ranges from -1 to 1. In the scope of the coordinate transformation, atten-

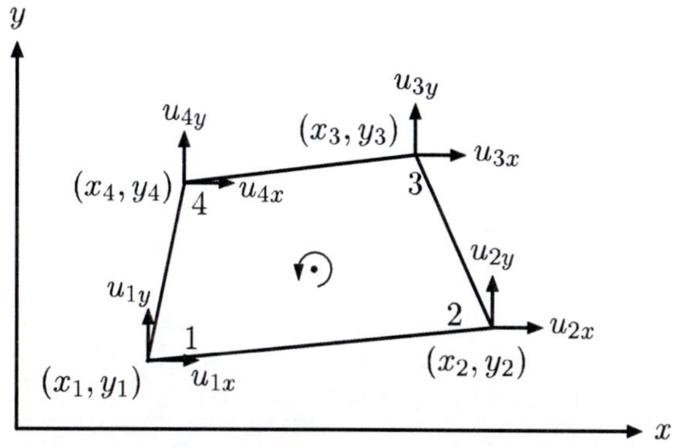

Fig. 3.2 Four-node planar element in the Cartesian space (x, y)

tion must be paid to the derivatives. For example, the derivative of the interpolation functions with respect to the x-coordinate is transformed in the following way:

$$\frac{\partial N_i}{\partial x} \rightarrow \frac{\partial N_i}{\partial \xi}\frac{\partial \xi}{\partial x} + \frac{\partial N_i}{\partial \eta}\frac{\partial \eta}{\partial x} . \tag{3.10}$$

Furthermore, the coordinate transformation requires that $\mathrm{d}V = t \times \mathrm{d}x\mathrm{d}y \rightarrow \mathrm{d}V' = t \times J\mathrm{d}\xi\mathrm{d}\eta$, where J is the Jacobian:

$$J = \left|\frac{\partial(x, y)}{\partial(\xi, \eta)}\right| = \frac{\partial x}{\partial \xi} \times \frac{\partial y}{\partial \eta} - \frac{\partial x}{\partial \eta} \times \frac{\partial y}{\partial \xi} . \tag{3.11}$$

In order to proceed, we need to have a closer look on the interpolation functions N_i. Under the assumption of a linear displacement field in parametric ξ-η space and four nodes per element, the following four interpolation functions can be easily derived, see [5] for details:

$$N_1(\xi, \eta) = \frac{1}{4}(1 - \xi - \eta + \xi\eta) = \frac{1}{4}(1 - \xi)(1 - \eta) , \tag{3.12}$$

$$N_2(\xi, \eta) = \frac{1}{4}(1 + \xi - \eta - \xi\eta) = \frac{1}{4}(1 + \xi)(1 - \eta) , \tag{3.13}$$

$$N_3(\xi, \eta) = \frac{1}{4}(1 + \xi + \eta + \xi\eta) = \frac{1}{4}(1 + \xi)(1 + \eta) , \tag{3.14}$$

$$N_4(\xi, \eta) = \frac{1}{4}(1 - \xi + \eta - \xi\eta) = \frac{1}{4}(1 - \xi)(1 + \eta) . \tag{3.15}$$

One may note that each N_i $(i = 1, 2, 3, 4)$ is unity when ξ and η assume coordinates of node i, but zero when ξ and η assume the coordinates of any other node. The graphical representation of the linear interpolation functions is shown in Fig. 3.3.

The derivatives with respect to the parametric coordinates can easily be obtained as:

$$\frac{\partial N_1(\xi, \eta)}{\partial \xi} = \frac{1}{4}(-1 + \eta) \quad ; \quad \frac{\partial N_1(\xi, \eta)}{\partial \eta} = \frac{1}{4}(-1 + \xi) , \tag{3.16}$$

$$\frac{\partial N_2(\xi, \eta)}{\partial \xi} = \frac{1}{4}(+1 - \eta) \quad ; \quad \frac{\partial N_2(\xi, \eta)}{\partial \eta} = \frac{1}{4}(-1 - \xi) , \tag{3.17}$$

$$\frac{\partial N_3(\xi, \eta)}{\partial \xi} = \frac{1}{4}(+1 + \eta) \quad ; \quad \frac{\partial N_3(\xi, \eta)}{\partial \eta} = \frac{1}{4}(+1 + \xi) , \tag{3.18}$$

$$\frac{\partial N_4(\xi, \eta)}{\partial \xi} = \frac{1}{4}(-1 - \eta) \quad ; \quad \frac{\partial N_4(\xi, \eta)}{\partial \eta} = \frac{1}{4}(+1 - \xi) . \tag{3.19}$$

To calculate the geometrical derivatives, let us assume the same interpolation for the global x- and y-coordinate as for the displacement (so-called *isoparametric* element formulation), i.e., $\overline{N}_i = N_i$:

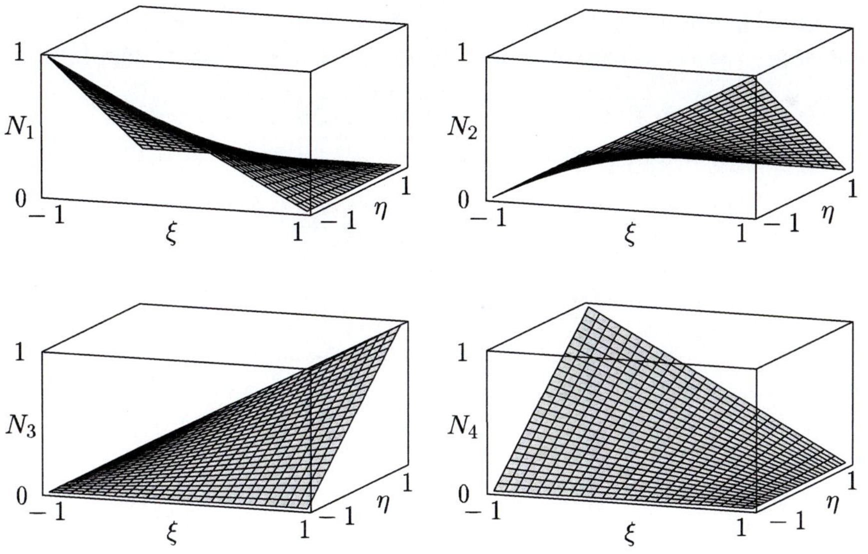

Fig. 3.3 Interpolation functions N_i $(i = 1, \ldots, 4)$ for a four-node planar element in parametric ξ-η space

$$\begin{aligned} x(\xi,\eta) &= \overline{N}_1(\xi,\eta)\times x_1 + \overline{N}_2(\xi,\eta)\times x_2 \\ &\quad + \overline{N}_3(\xi,\eta)\times x_3 + \overline{N}_4(\xi,\eta)\times x_4\,, \end{aligned} \tag{3.20}$$

$$\begin{aligned} y(\xi,\eta) &= \overline{N}_1(\xi,\eta)\times y_1 + \overline{N}_2(\xi,\eta)\times y_2 \\ &\quad + \overline{N}_3(\xi,\eta)\times y_3 + \overline{N}_4(\xi,\eta)\times y_4\,. \end{aligned} \tag{3.21}$$

Remark: the *global* coordinates of the nodes $1, \ldots, 4$ can be used for $x_1, \ldots, x_4$ and $y_1, \ldots, y_4$.

Thus, the geometrical derivatives can easily be obtained as:

$$\frac{\partial x}{\partial \xi} = \frac{1}{4}\Big((-1+\eta)x_1 + (1-\eta)x_2 + (1+\eta)x_3 + (-1-\eta)x_4\Big)\,, \tag{3.22}$$

$$\frac{\partial y}{\partial \xi} = \frac{1}{4}\Big((-1+\eta)y_1 + (1-\eta)y_2 + (1+\eta)y_3 + (-1-\eta)y_4\Big)\,, \tag{3.23}$$

$$\frac{\partial x}{\partial \eta} = \frac{1}{4}\Big((-1+\xi)x_1 + (-1-\xi)x_2 + (1+\xi)x_3 + (1-\xi)x_4\Big)\,, \tag{3.24}$$

$$\frac{\partial y}{\partial \eta} = \frac{1}{4}\Big((-1+\xi)y_1 + (-1-\xi)y_2 + (1+\xi)y_3 + (1-\xi)y_4\Big)\,. \tag{3.25}$$

The calculation of the derivatives of the interpolation functions (see Eq. (3.10)) requires, however, the geometrical derivatives of the natural coordinates (ξ, η) with respect to the physical coordinates (x, y). These relations can be easily obtained from Eqs. (3.22)–(3.25) under consideration of the following relationships [5]:

$$\frac{\partial \xi}{\partial x} = +\frac{1}{\frac{\partial x}{\partial \xi}\frac{\partial y}{\partial \eta} - \frac{\partial x}{\partial \eta}\frac{\partial y}{\partial \xi}} \times \frac{\partial y}{\partial \eta}, \tag{3.26}$$

$$\frac{\partial \xi}{\partial y} = -\frac{1}{\frac{\partial x}{\partial \xi}\frac{\partial y}{\partial \eta} - \frac{\partial x}{\partial \eta}\frac{\partial y}{\partial \xi}} \times \frac{\partial x}{\partial \eta}, \tag{3.27}$$

$$\frac{\partial \eta}{\partial x} = -\frac{1}{\frac{\partial x}{\partial \xi}\frac{\partial y}{\partial \eta} - \frac{\partial x}{\partial \eta}\frac{\partial y}{\partial \xi}} \times \frac{\partial y}{\partial \xi}, \tag{3.28}$$

$$\frac{\partial \eta}{\partial y} = +\frac{1}{\frac{\partial x}{\partial \xi}\frac{\partial y}{\partial \eta} - \frac{\partial x}{\partial \eta}\frac{\partial y}{\partial \xi}} \times \frac{\partial x}{\partial \xi}. \tag{3.29}$$

Based on the derived equations, the triple matrix product $\boldsymbol{BCB}^{\mathrm{T}}$ (see Eq. (3.7)) can be numerically calculated to obtain the stiffness matrix.

The integration in Eq. (3.7) is performed in commercial finite element packages based on Gauss-Legendre quadrature. For the domain integrals, one can write that

$$\int_{V_{\mathrm{e}}} f(x, y)\mathrm{d}V = \int_{V'_{\mathrm{e}}} f'(\xi, \eta) J \mathrm{d}V' = \int_{-1}^{1}\int_{-1}^{1} t f'(\xi, \eta) J \mathrm{d}\xi \mathrm{d}\eta = \sum_{i=1}^{n} t f'(\xi, \eta)_i J_i w_i \,, \tag{3.30}$$

where the Jacobian is $J = \frac{\partial x}{\partial \xi}\frac{\partial y}{\partial \eta} - \frac{\partial x}{\partial \eta}\frac{\partial y}{\partial \xi}$ (see Eq. 3.11), $(\xi, \eta)_i$ are the coordinates of the Gauss points and w_i are the corresponding weight factors. The location of the integration points and values of associated weights are given in Table 3.1. Furthermore, it should be highlighted that the thickness t in Eq. (3.30) is assumed constant.

Thus, we can write for the 2×2 integration indicated in Fig. 3.4:

Table 3.1 Integration rules for plane elasticity elements [3]

Points	ξ_i	η_i	Weight w_i	Error
1	0	0	4	$O(\xi^2)$
4	$\pm 1/\sqrt{3}$	$\pm 1/\sqrt{3}$	1	$O(\xi^4)$
9	0	0	$\frac{64}{81}$	$O(\xi^6)$
	0	$\pm\sqrt{6}$	$\frac{40}{81}$	
	$\pm\sqrt{6}$	0	$\frac{40}{81}$	
	$\pm\sqrt{6}$	$\pm\sqrt{6}$	$\frac{25}{81}$	

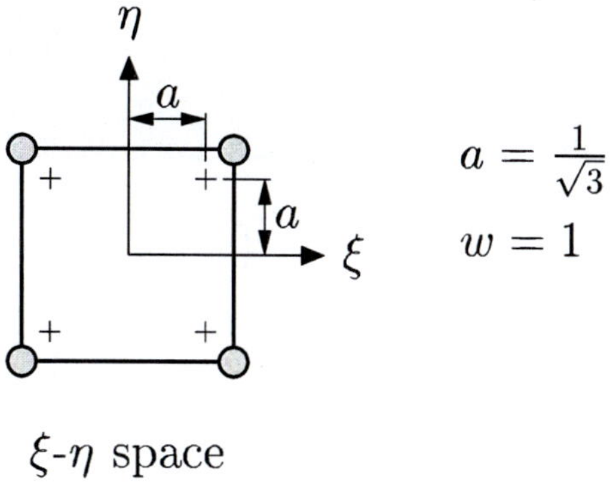

Fig. 3.4 Representation of a 2 × 2 integration for a plane elasticity element

$$
\begin{aligned}
\boldsymbol{K}^{\mathrm{e}} = \int_V (\boldsymbol{B}\boldsymbol{C}\boldsymbol{B}^{\mathrm{T}})\mathrm{d}V &= \boldsymbol{B}\boldsymbol{C}\boldsymbol{B}^{\mathrm{T}} J \times 1 \times t\Big|_{\left(-\frac{1}{\sqrt{3}},-\frac{1}{\sqrt{3}}\right)} \\
&+ \boldsymbol{B}\boldsymbol{C}\boldsymbol{B}^{\mathrm{T}} J \times 1 \times t\Big|_{\left(\frac{1}{\sqrt{3}},-\frac{1}{\sqrt{3}}\right)} \\
&+ \boldsymbol{B}\boldsymbol{C}\boldsymbol{B}^{\mathrm{T}} J \times 1 \times t\Big|_{\left(\frac{1}{\sqrt{3}},\frac{1}{\sqrt{3}}\right)} \\
&+ \boldsymbol{B}\boldsymbol{C}\boldsymbol{B}^{\mathrm{T}} J \times 1 \times t\Big|_{\left(-\frac{1}{\sqrt{3}},\frac{1}{\sqrt{3}}\right)} .
\end{aligned}
\tag{3.31}
$$

Let us summarize here the major steps which are required to calculate the elemental stiffness matrix.

❶ Introduce an elemental coordinate system (x, y).
❷ Express the coordinates (x_i, y_i) of the corner nodes i $(i = 1, \cdots, 4)$ in this elemental coordinate system.
❸ Calculate the partial derivatives of the old Cartesian (x, y) coordinates with respect to the new natural (ξ, η) coordinates, see Eqs. (3.22)–(3.25):

$$
\frac{\partial x}{\partial \xi} = x_{\xi} = \frac{1}{4}\Big((-1+\eta)x_1 + (1-\eta)x_2 + (1+\eta)x_3 + (-1-\eta)x_4\Big),
$$

$$
\vdots
$$

$$
\frac{\partial y}{\partial \eta} = y_{\eta} = \frac{1}{4}\Big((-1+\xi)y_1 + (-1-\xi)y_2 + (1+\xi)y_3 + (1-\xi)y_4\Big).
$$

❹ Calculate the partial derivatives of the new natural (ξ, η) coordinates with respect to the old Cartesian (x, y) coordinates, see Eqs. (3.26)–(3.29):

$$
\frac{\partial \xi}{\partial x} = +\frac{1}{\frac{\partial x}{\partial \xi}\frac{\partial y}{\partial \eta} - \frac{\partial x}{\partial \eta}\frac{\partial y}{\partial \xi}} \times \frac{\partial y}{\partial \eta}, \quad \frac{\partial \xi}{\partial y} = -\frac{1}{\frac{\partial x}{\partial \xi}\frac{\partial y}{\partial \eta} - \frac{\partial x}{\partial \eta}\frac{\partial y}{\partial \xi}} \times \frac{\partial x}{\partial \eta},
$$

$$\frac{\partial \eta}{\partial x} = -\frac{1}{\frac{\partial x}{\partial \xi}\frac{\partial y}{\partial \eta} - \frac{\partial x}{\partial \eta}\frac{\partial y}{\partial \xi}} \times \frac{\partial y}{\partial \xi}, \frac{\partial \eta}{\partial y} = +\frac{1}{\frac{\partial x}{\partial \xi}\frac{\partial y}{\partial \eta} - \frac{\partial x}{\partial \eta}\frac{\partial y}{\partial \xi}} \times \frac{\partial x}{\partial \xi}.$$

❺ Calculate the $\boldsymbol{B}$-matrix and its transposed, see Eq. (3.9):

$$\boldsymbol{B}^{\mathrm{T}} = \left[\begin{array}{cc|cc|cc|cc} \frac{\partial N_1}{\partial x} & 0 & \frac{\partial N_2}{\partial x} & 0 & \frac{\partial N_3}{\partial x} & 0 & \frac{\partial N_4}{\partial x} & 0 \\ 0 & \frac{\partial N_1}{\partial y} & 0 & \frac{\partial N_2}{\partial y} & 0 & \frac{\partial N_3}{\partial y} & 0 & \frac{\partial N_4}{\partial y} \\ \frac{\partial N_1}{\partial y} & \frac{\partial N_1}{\partial x} & \frac{\partial N_2}{\partial y} & \frac{\partial N_2}{\partial x} & \frac{\partial N_3}{\partial y} & \frac{\partial N_3}{\partial x} & \frac{\partial N_4}{\partial y} & \frac{\partial N_4}{\partial x} \end{array}\right],$$

where the partial derivatives are $\frac{\partial N_1(\xi,\eta)}{\partial x} = \frac{\partial N_1}{\partial \xi}\frac{\partial \xi}{\partial x} + \frac{\partial N_1}{\partial \eta}\frac{\partial \eta}{\partial x}, \ldots$ and the derivatives of the interpolation functions are given in Eqs. (3.16)–(3.19), i.e., $\frac{\partial N_1}{\partial \xi} = \frac{1}{4}(-1+\eta), \ldots$

❻ Calculate the triple matrix product $\boldsymbol{BCB}^{\mathrm{T}}$, where the elasticity matrix $\boldsymbol{C}$ is given by Eqs. (3.1) and (3.4).

❼ Perform the numerical integration based on a 2×2 integration rule:

$$\int_V (\boldsymbol{BCB}^{\mathrm{T}})\mathrm{d}V = \boldsymbol{BCB}^{\mathrm{T}} J \times 1 \times t\Big|_{\left(-\frac{1}{\sqrt{3}},-\frac{1}{\sqrt{3}}\right)}$$
$$+ \boldsymbol{BCB}^{\mathrm{T}} J \times 1 \times t\Big|_{\left(\frac{1}{\sqrt{3}},-\frac{1}{\sqrt{3}}\right)} + \boldsymbol{BCB}^{\mathrm{T}} J \times 1 \times t\Big|_{\left(\frac{1}{\sqrt{3}},\frac{1}{\sqrt{3}}\right)}$$
$$+ \boldsymbol{BCB}^{\mathrm{T}} J \times 1 \times t\Big|_{\left(-\frac{1}{\sqrt{3}},\frac{1}{\sqrt{3}}\right)}.$$

❽ $\boldsymbol{K}^{\mathrm{e}}$ obtained.

3.1.2 Elements with Three Nodes

Let us consider in the following the three-node planar element as shown in Fig. 3.5. Important for the following derivations is that the node numbering is counterclockwise. Furthermore, all the following derivations are restricted to a linear three-node element under the plane stress or plane strain assumption.[1]

The general expression for the elemental stiffness matrix $\boldsymbol{K}^{\mathrm{e}}$ of a three-node plane elasticity element with elasticity matrix $\boldsymbol{C}$ can be expressed as

$$\boldsymbol{K}^{\mathrm{e}} = \int_V \underbrace{(\boldsymbol{\mathcal{L}}_1 \boldsymbol{N}^{\mathrm{T}})^{\mathrm{T}}}_{\boldsymbol{B}} \boldsymbol{C} \underbrace{(\boldsymbol{\mathcal{L}}_1 \boldsymbol{N}^{\mathrm{T}})}_{\boldsymbol{B}^{\mathrm{T}}} \mathrm{d}V, \tag{3.32}$$

where the $\boldsymbol{B}$-matrix contains the derivatives of the interpolation functions N_i:

[1] The plane stress element is known in literature as the constant strain triangle or CST element.

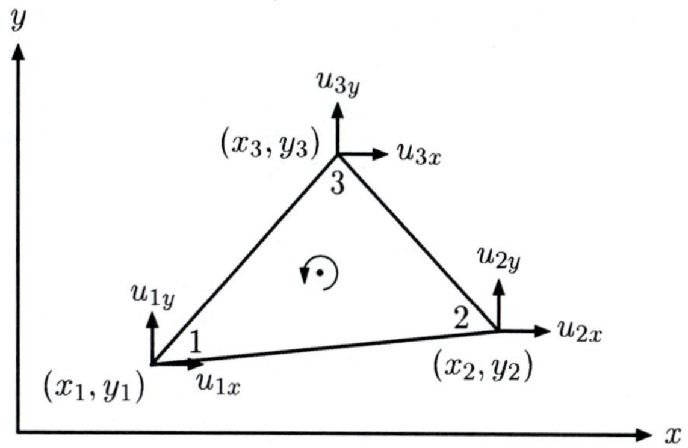

Fig. 3.5 Three-node planar element in the Cartesian space (x, y)

$$\boldsymbol{B} = \left(\mathcal{L}_1 \boldsymbol{N}^{\mathrm{T}}\right)^{\mathrm{T}} = \left(\begin{bmatrix} \frac{\partial}{\partial x} & 0 \\ 0 & \frac{\partial}{\partial y} \\ \frac{\partial}{\partial y} & \frac{\partial}{\partial x} \end{bmatrix} \left[\begin{array}{cc|cc|cc} N_1 & 0 & N_2 & 0 & N_3 & 0 \\ 0 & N_1 & 0 & N_2 & 0 & N_3 \end{array} \right] \right)^{\mathrm{T}} \tag{3.33}$$

$$= \left(\left[\begin{array}{cc|cc|cc} \frac{\partial N_1(x,y)}{\partial x} & 0 & \frac{\partial N_2(x,y)}{\partial x} & 0 & \frac{\partial N_3(x,y)}{\partial x} & 0 \\ 0 & \frac{\partial N_1(x,y)}{\partial y} & 0 & \frac{\partial N_2(x,y)}{\partial y} & 0 & \frac{\partial N_3(x,y)}{\partial y} \\ \frac{\partial N_1(x,y)}{\partial y} & \frac{\partial N_1(x,y)}{\partial x} & \frac{\partial N_2(x,y)}{\partial y} & \frac{\partial N_2(x,y)}{\partial x} & \frac{\partial N_3(x,y)}{\partial y} & \frac{\partial N_3(x,y)}{\partial x} \end{array} \right] \right)^{\mathrm{T}}, \tag{3.34}$$

which is a (6×3)-matrix. The transposed, i.e., $\boldsymbol{B}^{\mathrm{T}} = \mathcal{L}_1 \boldsymbol{N}^{\mathrm{T}}$, is thus a (3×6)-matrix. The triple matrix multiplication gives after integration the stiffness matrix with a dimension of 6×6.

The integration for the element matrix given in Eq. (3.32) is again approximated by numerical integration in commercial finite element packages. To this end, the coordinates (x, y) are transformed to the natural coordinates $(\xi, \eta(, \zeta))$. In the scope of the coordinate transformation, attention must be paid to the derivatives. For example, the derivative of the interpolation functions with respect to the x-coordinate is transformed in to the following way:

$$\frac{\partial N_i}{\partial x} \rightarrow \frac{\partial N_i}{\partial \xi}\frac{\partial \xi}{\partial x} + \frac{\partial N_i}{\partial \eta}\frac{\partial \eta}{\partial x}. \tag{3.35}$$

Furthermore, the coordinate transformation requires that $\mathrm{d}V = t \times \mathrm{d}x\mathrm{d}y \rightarrow \mathrm{d}V' = t \times J \mathrm{d}\xi\mathrm{d}\eta$, where J is the Jacobian:

$$J = \left| \frac{\partial(x, y)}{\partial(\xi, \eta)} \right| = \frac{\partial x}{\partial \xi} \times \frac{\partial y}{\partial \eta} - \frac{\partial x}{\partial \eta} \times \frac{\partial y}{\partial \xi}. \tag{3.36}$$

In the case of triangular elements, the so-called triangular coordinates[2] can be stated for any point P (see Fig. 3.6) as:

$$\xi = \frac{A_1}{A}, \ \eta = \frac{A_2}{A}, \ \zeta = \frac{A_3}{A}. \tag{3.37}$$

The area A_1, for example, is obtained by connecting the two nodes other than 1 with point P (see Fig. 3.6a) and A is the total area of the triangle. Furthermore, it should be noted that each triangular coordinate ranges between 0 and 1, see Fig. 3.6b and c.

Now, the three nodes of the triangle shown in Fig. 3.5 can be expressed based on these triangular coordinates as (see the graphical representation of the three-node element in the natural space ξ-η in Fig. 3.7):

$$\text{Node 1}: \ \xi = 1, \eta = 0, \zeta = 0\,, \tag{3.38}$$

$$\text{Node 2}: \ \xi = 0, \eta = 1, \zeta = 0\,, \tag{3.39}$$

$$\text{Node 3}: \ \xi = 0, \eta = 0, \zeta = 1\,. \tag{3.40}$$

This can be generalized to express the coordinate of the reference point $P(x_P, y_P)$ in triangular coordinates as:

$$x_P = x_1\xi + x_2\eta + x_3\zeta\,, \tag{3.41}$$

$$y_P = y_1\xi + y_2\eta + y_3\zeta\,, \tag{3.42}$$

where (x_i, y_i) are the Cartesian coordinates of node i. Furthermore, we can deduct from the definition of the triangular coordinates (see Eq. (3.37)) that:

$$\xi + \eta + \zeta = \frac{A_1}{A} + \frac{A_2}{A} + \frac{A_3}{A} = \frac{1}{A}(A_1 + A_2 + A_3) = 1\,. \tag{3.43}$$

Introducing this relationship in Eqs. (3.41)–(3.42), we get

$$x_P = x_1\xi + x_2\eta + x_3\underbrace{(1 - \xi - \eta)}_{\zeta}\,, \tag{3.44}$$

$$y_P = y_1\xi + y_2\eta + y_3\underbrace{(1 - \xi - \eta)}_{\zeta}\,, \tag{3.45}$$

and a comparison with the general nodal approach, i.e.,

$$x(\xi, \eta) = \sum_{i=1}^{3} N_i(\xi, \eta)x_i \ ; \ \ y(\xi, \eta) = \sum_{i=1}^{3} N_i(\xi, \eta)y_i\,, \tag{3.46}$$

[2] Alternatively named as area coordinates or barycentric coordinates.

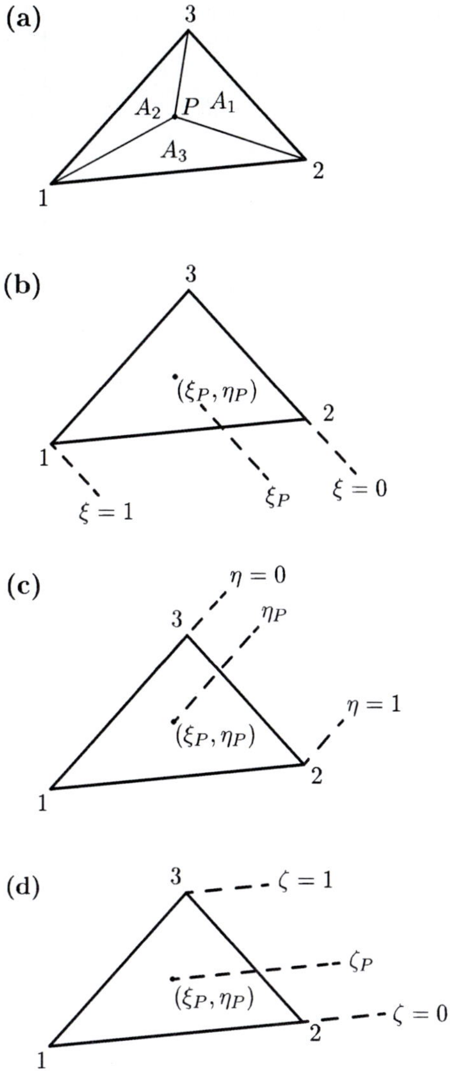

Fig. 3.6 Graphical representation of triangular coordinates: **a** definition of subareas A_i, **b** range of coordinate ξ, **c** range of coordinate η, **d** range of coordinate ζ

allows under the assumption of an *isoparametric* element formulation the deduction of the three interpolation functions in triangular coordinates as:

$$N_1(\xi, \eta) = \xi \,, \tag{3.47}$$

$$N_2(\xi, \eta) = \eta \,, \tag{3.48}$$

$$N_3(\xi, \eta) = 1 - \xi - \eta \; (= \zeta) \,. \tag{3.49}$$

Equations (3.43) and (3.41)–(3.42) can be combined in matrix notation to obtain:

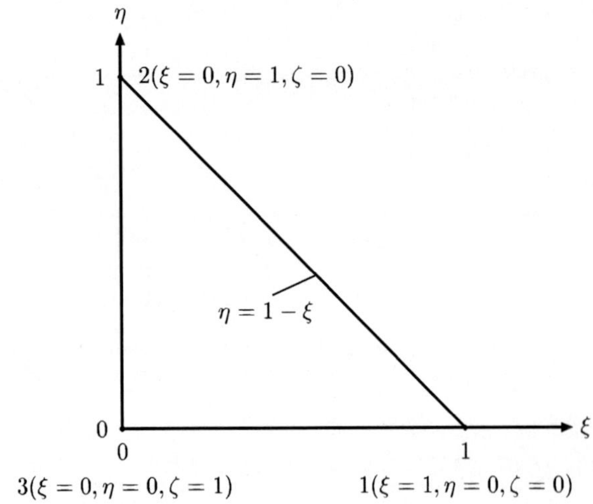

Fig. 3.7 Three-node planar element in the natural space (ξ, η)

$$\begin{bmatrix} 1 \\ x_P \\ y_P \end{bmatrix} = \begin{bmatrix} 1 & 1 & 1 \\ x_1 & x_2 & x_3 \\ y_1 & y_2 & y_3 \end{bmatrix} \begin{bmatrix} \xi \\ \eta \\ \zeta \end{bmatrix}. \tag{3.50}$$

The last equation can be solved for the triangular coordinates as a function of the Cartesian coordinates of the reference point P[3]:

$$\begin{bmatrix} \xi \\ \eta \\ \zeta \end{bmatrix} = \frac{1}{x_1(y_2 - y_3) + x_2(y_3 - y_1) + x_3(y_1 - y_2)} \times$$

$$\times \begin{bmatrix} x_2y_3 - x_3y_2 & y_2 - y_3 & x_3 - x_2 \\ x_3y_1 - x_1y_3 & y_3 - y_1 & x_1 - x_3 \\ x_1y_2 - x_2y_2 & y_1 - y_2 & x_2 - x_1 \end{bmatrix} \begin{bmatrix} 1 \\ x_P \\ y_P \end{bmatrix} \tag{3.51}$$

$$= \frac{1}{2A} \begin{bmatrix} x_2y_3 - x_3y_2 & y_2 - y_3 & x_3 - x_2 \\ x_3y_1 - x_1y_3 & y_3 - y_1 & x_1 - x_3 \\ x_1y_2 - x_2y_2 & y_1 - y_2 & x_2 - x_1 \end{bmatrix} \begin{bmatrix} 1 \\ x_P \\ y_P \end{bmatrix}. \tag{3.52}$$

The partial derivatives of the interpolation functions (expressed in triangular coordinates) with respect to the Cartesian coordinates can be written under consideration of the chain rule as follows:

$$\frac{\partial N_1(\xi, \eta)}{\partial x} = \frac{\partial N_1}{\partial \xi}\frac{\partial \xi}{\partial x} + \frac{\partial N_1}{\partial \eta}\frac{\partial \eta}{\partial x}, \tag{3.53}$$

[3] If we skip the index 'P', then we simply have the relationship between the triangular and the Cartesian coordinates.

$$\frac{\partial N_1(\xi,\eta)}{\partial y} = \frac{\partial N_1}{\partial \xi}\frac{\partial \xi}{\partial y} + \frac{\partial N_1}{\partial \eta}\frac{\partial \eta}{\partial y}, \tag{3.54}$$

$$\vdots \tag{3.55}$$

$$\frac{\partial N_3(\xi,\eta)}{\partial x} = \frac{\partial N_3}{\partial \xi}\frac{\partial \xi}{\partial x} + \frac{\partial N_3}{\partial \eta}\frac{\partial \eta}{\partial x}, \tag{3.56}$$

$$\frac{\partial N_3(\xi,\eta)}{\partial y} = \frac{\partial N_3}{\partial \xi}\frac{\partial \xi}{\partial y} + \frac{\partial N_3}{\partial \eta}\frac{\partial \eta}{\partial y}. \tag{3.57}$$

Partial derivatives of the interpolation functions with respect to the triangular coordinates can be easily obtained from Eqs. (3.47)–(3.49) as:

$$\frac{\partial N_1(\xi,\eta)}{\partial \xi} = 1, \qquad \frac{\partial N_1(\xi,\eta)}{\partial \eta} = 0, \tag{3.58}$$

$$\frac{\partial N_2(\xi,\eta)}{\partial \xi} = 0, \qquad \frac{\partial N_2(\xi,\eta)}{\partial \eta} = 1, \tag{3.59}$$

$$\frac{\partial N_3(\xi,\eta)}{\partial \xi} = -1, \qquad \frac{\partial N_3(\xi,\eta)}{\partial \eta} = -1. \tag{3.60}$$

The geometrical derivatives result from Eq. (3.52) as:

$$\frac{\partial \xi(x,y)}{\partial x} = \frac{y_2 - y_3}{2A}, \qquad \frac{\partial \xi(x,y)}{\partial y} = \frac{x_3 - x_2}{2A}, \tag{3.61}$$

$$\frac{\partial \eta(x,y)}{\partial x} = \frac{y_3 - y_1}{2A}, \qquad \frac{\partial \eta(x,y)}{\partial y} = \frac{x_1 - x_3}{2A}. \tag{3.62}$$

Thus, we can finally write the partial derivatives as follows:

$$\frac{\partial N_1(\xi,\eta)}{\partial x} = \frac{y_2 - y_3}{2A}, \qquad \frac{\partial N_1(\xi,\eta)}{\partial y} = \frac{x_3 - x_2}{2A}, \tag{3.63}$$

$$\frac{\partial N_2(\xi,\eta)}{\partial x} = \frac{y_3 - y_1}{2A}, \qquad \frac{\partial N_2(\xi,\eta)}{\partial y} = \frac{x_1 - x_3}{2A}, \tag{3.64}$$

$$\frac{\partial N_3(\xi,\eta)}{\partial x} = \frac{y_1 - y_2}{2A}, \qquad \frac{\partial N_3(\xi,\eta)}{\partial y} = \frac{x_2 - x_1}{2A}, \tag{3.65}$$

which results in the following $\boldsymbol{B}$-matrix:

$$\boldsymbol{B} = \left(\frac{1}{2A} \left[\begin{array}{cc|cc|cc} y_2 - y_3 & 0 & y_3 - y_1 & 0 & y_1 - y_2 & 0 \\ 0 & x_3 - x_2 & 0 & x_1 - x_3 & 0 & x_2 - x_1 \\ x_3 - x_2 & y_2 - y_3 & x_1 - x_3 & y_3 - y_1 & x_2 - x_1 & y_1 - y_2 \end{array} \right] \right)^{\mathrm{T}}. \tag{3.66}$$

Table 3.2 Symmetric quadrature for the unit triangle (see Fig. 3.7). Adapted from [1]

$$\int_{\eta=0}^{1}\int_{\xi=0}^{1-\eta} f(\xi,\eta)\mathrm{d}\xi\mathrm{d}\eta = \sum_{i=1}^{n} f(\xi_i,\eta_i)w_i$$

No. Points n	P	Int. Point i	Coordinate ξ_i	Coordinate η_i	Weights w_i
1	1	1	1/3	1/3	1/2
3	2	1	1/6	1/6	1/6
		2	2/3	1/6	1/6
		3	1/6	2/3	1/6
4	3	1	1/3	1/3	−9/32
		2	3/5	1/5	25/96
		3	1/5	3/5	25/96
		4	1/5	1/5	25/96
7	4	1	0	0	1/40
		2	1/2	0	1/15
		3	1	0	1/40
		4	1/2	1/2	1/15
		5	0	1	1/40
		6	0	1/2	1/15
		7	1/3	1/3	9/40

P = Degree of Polynomial for exact integration

To calculate the Jacobian according to Eq. (3.36), we can derive from Eq. (3.50) the following geometrical derivatives of the Cartesian coordinates with respect to the area coordinates:

$$\frac{\partial x(\xi,\eta)}{\partial \xi} = x_1 - x_3\,, \qquad \frac{\partial x(\xi,\eta)}{\partial \eta} = x_2 - x_3\,, \tag{3.67}$$

$$\frac{\partial y(\xi,\eta)}{\partial \xi} = y_1 - y_3\,, \qquad \frac{\partial y(\xi,\eta)}{\partial \eta} = y_2 - y_3\,. \tag{3.68}$$

It should be noted here that the evaluation of the Jacobian gives here $J = 2A$. The integration based on triangular coordinates should be handled with care. In the case of a triangle with a constant thickness, the following integration rules can be applied for triangles with a constant Jacobian (e.g., straight sides and flat faces)[4]:

$$\int_A \xi^i \eta^j \zeta^k \mathrm{d}x\mathrm{d}y = \frac{i!j!k!}{(i+j+k+2)!} \times 2A\,, \tag{3.69}$$

$$\int_A \xi^i \eta^j \zeta^k \mathrm{d}\xi\mathrm{d}\eta = \frac{i!j!k!}{(i+j+k+2)!}\,, \tag{3.70}$$

for $i \geq 0$, $j \geq 0$, and $k \geq 0$. Worth considering are the following special cases:

[4] Remember: $0! = 1$, $1! = 1$, $2! = 2$, and $3! = 6$.

$$\int_A 1\mathrm{d}x\mathrm{d}y = \frac{1}{2} \times 2A = A\,, \tag{3.71}$$

$$\int_A 1\mathrm{d}\xi\mathrm{d}\eta = \frac{1}{2}, \tag{3.72}$$

or

$$\int_A \xi\mathrm{d}x\mathrm{d}y = \frac{1}{6} \times 2A = \frac{A}{3}, \tag{3.73}$$

$$\int_A \xi\mathrm{d}\xi\mathrm{d}\eta = \frac{1}{6}. \tag{3.74}$$

The numerical integration over the area of unit triangles can be performed based on the symmetric quadrature rule given in Table 3.2. The locations of the corresponding integration points are illustrated in Fig. 3.8.

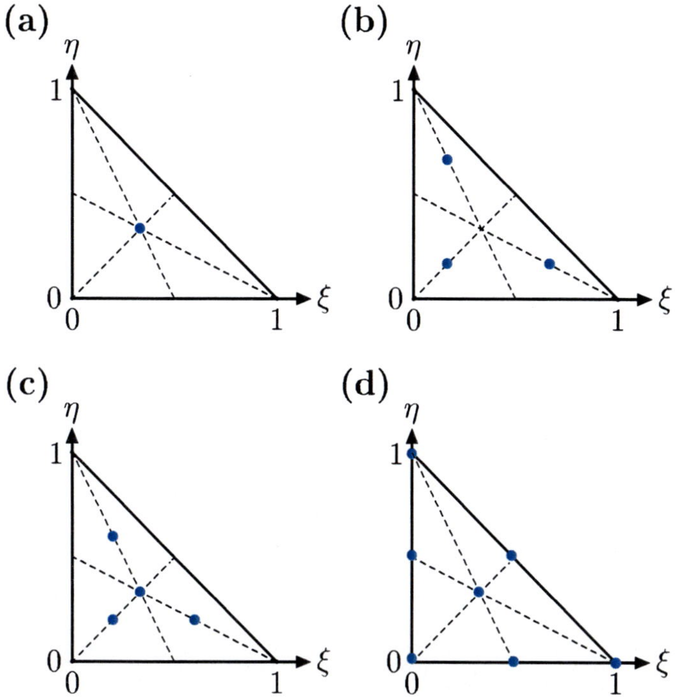

Fig. 3.8 Symmetric quadrature locations for unit triangle: **a** $n = 1$, **b** $n = 3$, **c** $n = 4$, **d** $n = 7$. Adapted from [1]

3.1.3 Maxima Modules for Stiffness Matrices

3.1.4 Maxima Examples to Calculate the Stiffness Matrix

3.1 Example: Two-dimensional rectangular element under plane stress condition

Given is a regular two-dimensional element as shown in Fig. 3.9.

Derive the general expression for the elemental stiffness matrix under the plane stress assumption. Furthermore, it can be assumed that the thickness t is constant. Use first analytical integration to obtain the stiffness matrix and then compare the results with a 2×2 Gauss-Legendre integration rule.

3.1 Solution

The following Module 3.1 shows the entire wxMaxima code for the derivation of $\boldsymbol{K}^{\mathrm{e}}$ based on analytical integration.

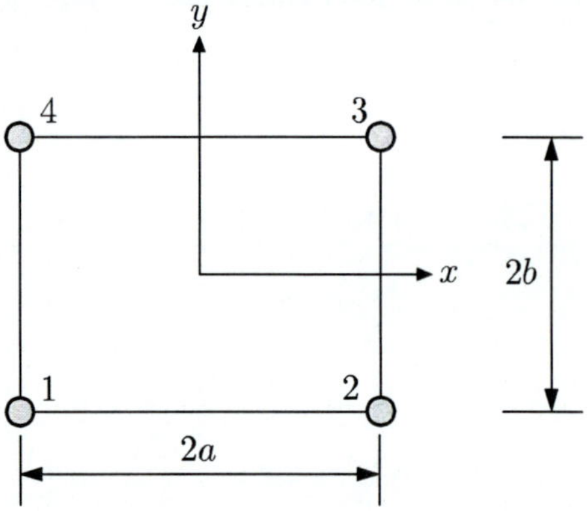

Fig. 3.9 Two-dimensional rectangular element under plane stress condition

```
→       kill(all);
        load("../Library/my_funs.mac")$
        ratprint : false$

(% o0)  done
```

Definition of Plane Stress Stiffness: Nodal Coordinates "ncoor", Material Properties "mat", Number of Element Nodes "nen", Thickness Function "thickness"

```
→       ncoor : [[-a,-b],[a,-b],[a,b],[-a,b]]$

        mat : [E, v]$

        thickness(x,y) := t$

        type : "plane stress"$
        nen : 4$
```

Maxima Calculations and Output

```
→       Ke :
        k_el_2D_plane_analytical(ncoor, mat, nen, type)$

        print(" ")$
        print("Element stiffness matrix:")$
        coeff : E*t/(1-v^2)$
        print(" ",coeff,factor(ratsimp(Ke/coeff)))$
```

Element Stiffness Matrix:

$$\frac{Et}{1-v^2}$$

$$\begin{bmatrix}
-\frac{a^2v-2b^2-a^2}{6ab} & \frac{v+1}{8} & -\frac{a^2v+4b^2-a^2}{12ab} & \frac{3v-1}{8} & \frac{a^2v-2b^2-a^2}{12ab} & -\frac{v+1}{8} & \cdots \\
\frac{v+1}{8} & -\frac{b^2v-b^2-2a^2}{6ab} & -\frac{3v-1}{8} & \frac{b^2v-b^2+a^2}{6ab} & -\frac{v+1}{8} & \frac{b^2v-b^2-2a^2}{12ab} & \cdots \\
-\frac{a^2v+4b^2-a^2}{12ab} & -\frac{3v-1}{8} & -\frac{a^2v-2b^2-a^2}{6ab} & -\frac{v+1}{8} & \frac{a^2v+b^2-a^2}{6ab} & \frac{3v-1}{8} & \cdots \\
\frac{3v-1}{8} & \frac{b^2v-b^2+a^2}{6ab} & -\frac{v+1}{8} & -\frac{b^2v-b^2-2a^2}{6ab} & -\frac{3v-1}{8} & -\frac{b^2v-b^2+4a^2}{12ab} & \cdots \\
\frac{a^2v-2b^2-a^2}{12ab} & -\frac{v+1}{8} & \frac{a^2v+b^2-a^2}{6ab} & -\frac{3v-1}{8} & -\frac{a^2v-2b^2-a^2}{6ab} & \frac{v+1}{8} & \cdots \\
-\frac{v+1}{8} & \frac{b^2v-b^2-2a^2}{12ab} & \frac{3v-1}{8} & -\frac{b^2v-b^2+4a^2}{12ab} & \frac{v+1}{8} & -\frac{b^2v-b^2-2a^2}{6ab} & \cdots \\
\frac{a^2v+b^2-a^2}{6ab} & \frac{3v-1}{8} & \frac{a^2v-2b^2-a^2}{12ab} & \frac{v+1}{8} & -\frac{a^2v+4b^2-a^2}{12ab} & -\frac{3v-1}{8} & \cdots \\
-\frac{3v-1}{8} & -\frac{b^2v-b^2+4a^2}{12ab} & \frac{v+1}{8} & \frac{b^2v-b^2-2a^2}{12ab} & \frac{3v-1}{8} & \frac{b^2v-b^2+a^2}{6ab} & \cdots
\end{bmatrix}$$

Module 3.1: Stiffness matrix for a two-dimensional rectangular element ($2a \times 2b$) under plane stress condition (analytical integration). The complete stiffness matrix is shown in Eq. (3.75)

$$
\boldsymbol{K}^{\mathrm{e}} = \frac{Et}{1-v^2} \times
$$

$$
\begin{bmatrix}
-\frac{a^2v-2b^2-a^2}{6ab} & \frac{v+1}{8} & -\frac{a^2v+4b^2-a^2}{12ab} & \frac{3v-1}{8} & \frac{a^2v-2b^2-a^2}{12ab} & -\frac{v+1}{8} & \frac{a^2v+b^2-a^2}{6ab} & -\frac{3v-1}{8} \\
\frac{v+1}{8} & -\frac{b^2v-b^2-2a^2}{6ab} & -\frac{3v-1}{8} & \frac{b^2v-b^2+a^2}{6ab} & -\frac{v+1}{8} & \frac{b^2v-b^2-2a^2}{12ab} & \frac{3v-1}{8} & -\frac{b^2v-b^2+4a^2}{12ab} \\
-\frac{a^2v+4b^2-a^2}{12ab} & -\frac{3v-1}{8} & -\frac{a^2v-2b^2-a^2}{6ab} & -\frac{v+1}{8} & \frac{a^2v+b^2-a^2}{6ab} & \frac{3v-1}{8} & \frac{a^2v-2b^2-a^2}{12ab} & \frac{v+1}{8} \\
\frac{3v-1}{8} & \frac{b^2v-b^2+a^2}{6ab} & -\frac{v+1}{8} & -\frac{b^2v-b^2-2a^2}{6ab} & -\frac{3v-1}{8} & -\frac{b^2v-b^2+4a^2}{12ab} & \frac{v+1}{8} & \frac{b^2v-b^2-2a^2}{12ab} \\
\frac{a^2v-2b^2-a^2}{12ab} & -\frac{v+1}{8} & \frac{a^2v+b^2-a^2}{6ab} & -\frac{3v-1}{8} & -\frac{a^2v-2b^2-a^2}{6ab} & \frac{v+1}{8} & -\frac{a^2v+4b^2-a^2}{12ab} & \frac{3v-1}{8} \\
-\frac{v+1}{8} & \frac{b^2v-b^2-2a^2}{12ab} & \frac{3v-1}{8} & -\frac{b^2v-b^2+4a^2}{12ab} & \frac{v+1}{8} & -\frac{b^2v-b^2-2a^2}{6ab} & -\frac{3v-1}{8} & \frac{b^2v-b^2+a^2}{6ab} \\
\frac{a^2v+b^2-a^2}{6ab} & \frac{3v-1}{8} & \frac{a^2v-2b^2-a^2}{12ab} & \frac{v+1}{8} & -\frac{a^2v+4b^2-a^2}{12ab} & -\frac{3v-1}{8} & -\frac{a^2v-2b^2-a^2}{6ab} & -\frac{v+1}{8} \\
-\frac{3v-1}{8} & -\frac{b^2v-b^2+4a^2}{12ab} & \frac{v+1}{8} & \frac{b^2v-b^2-2a^2}{12ab} & \frac{3v-1}{8} & \frac{b^2v-b^2+a^2}{6ab} & -\frac{v+1}{8} & -\frac{b^2v-b^2-2a^2}{6ab}
\end{bmatrix}. \tag{3.75}
$$

The following Module 3.2 shows the entire wxMaxima code for the derivation of $\boldsymbol{K}^{\mathrm{e}}$ based on a 2×2 Gauss-Legendre integration rule

```
-->  kill(all);
     load("../Library/my_funs.mac")$
     ratprint : false$

(% o0)  done
```

Definition of Plane Stress Stiffness: Nodal Coordinates "ncoor", Material Properties "mat", Number of Element Nodes "nen", Thickness Function "thickness", Number of Quadrature Points "nqp"

```
-->  ncoor : [[-a,-b],[a,-b],[a,b],[-a,b]]$

     mat : [E, v]$

     thickness(x,y) := t$

     type : "plane stress"$
     nen : 4$
     nqp : 4$
```

Maxima Calculations and Output

```
-->  Ke :
     k_el_2D_plane_numerical(ncoor, mat, nqp, nen, type)$

     print(" ")$
     print("Element stiffness matrix:")$
     coeff : E*t/(1-v^2)$
     print(" ",coeff,factor(ratsimp(Ke/coeff)))$
```

Element Stiffness Matrix:

$$\frac{Et}{1-v^2}$$

$$\begin{bmatrix}
-\frac{a^2v-2b^2-a^2}{6ab} & \frac{v+1}{8} & -\frac{a^2v+4b^2-a^2}{12ab} & \frac{3v-1}{8} & \frac{a^2v-2b^2-a^2}{12ab} & -\frac{v+1}{8} & \cdots \\
\frac{v+1}{8} & -\frac{b^2v-b^2-2a^2}{6ab} & -\frac{3v-1}{8} & \frac{b^2v-b^2+a^2}{6ab} & -\frac{v+1}{8} & \frac{b^2v-b^2-2a^2}{12ab} & \cdots \\
-\frac{a^2v+4b^2-a^2}{12ab} & -\frac{3v-1}{8} & -\frac{a^2v-2b^2-a^2}{6ab} & -\frac{v+1}{8} & \frac{a^2v+b^2-a^2}{6ab} & \cdots & \\
\frac{3v-1}{8} & \frac{b^2v-b^2+a^2}{6ab} & -\frac{v+1}{8} & -\frac{b^2v-b^2-2a^2}{6ab} & -\frac{3v-1}{8} & -\frac{b^2v-b^2+4a^2}{12ab} & \cdots \\
\frac{a^2v-2b^2-a^2}{12ab} & -\frac{v+1}{8} & \frac{a^2v+b^2-a^2}{6ab} & -\frac{3v-1}{8} & -\frac{a^2v-2b^2-a^2}{6ab} & \cdots & \\
-\frac{v+1}{8} & \frac{b^2v-b^2-2a^2}{12ab} & \frac{3v-1}{8} & -\frac{b^2v-b^2+4a^2}{12ab} & \frac{v+1}{8} & -\frac{b^2v-b^2-2a^2}{6ab} & \cdots \\
\frac{a^2v+b^2-a^2}{6ab} & \frac{3v-1}{8} & \frac{a^2v-2b^2-a^2}{12ab} & \frac{v+1}{8} & -\frac{a^2v+4b^2-a^2}{12ab} & \cdots & \\
-\frac{3v-1}{8} & -\frac{b^2v-b^2+4a^2}{12ab} & \frac{v+1}{8} & \frac{b^2v-b^2-2a^2}{12ab} & \frac{3v-1}{8} & \frac{b^2v-b^2+a^2}{6ab} & \cdots
\end{bmatrix}$$

Module 3.2: Stiffness matrix for a two-dimensional rectangular element ($2a \times 2b$) under plane stress condition (2×2 numerical integration). The complete stiffness matrix is shown in Eq. (3.76)

$$
\boldsymbol{K}^{\mathrm{e}} = \frac{Et}{1-v^2} \times
\begin{bmatrix}
-\frac{a^2v-2b^2-a^2}{6ab} & \frac{v+1}{8} & -\frac{a^2v+4b^2-a^2}{12ab} & \frac{3v-1}{8} & \frac{a^2v-2b^2-a^2}{12ab} & -\frac{v+1}{8} & \frac{a^2v+b^2-a^2}{6ab} & -\frac{3v-1}{8} \\
\frac{v+1}{8} & -\frac{b^2v-b^2-2a^2}{6ab} & -\frac{3v-1}{8} & \frac{b^2v-b^2+a^2}{6ab} & -\frac{v+1}{8} & \frac{b^2v-b^2-2a^2}{12ab} & \frac{3v-1}{8} & -\frac{b^2v-b^2+4a^2}{12ab} \\
-\frac{a^2v+4b^2-a^2}{12ab} & -\frac{3v-1}{8} & -\frac{a^2v-2b^2-a^2}{6ab} & -\frac{v+1}{8} & \frac{a^2v+b^2-a^2}{6ab} & \frac{3v-1}{8} & \frac{a^2v-2b^2-a^2}{12ab} & \frac{v+1}{8} \\
\frac{3v-1}{8} & \frac{b^2v-b^2+a^2}{6ab} & -\frac{v+1}{8} & -\frac{b^2v-b^2-2a^2}{6ab} & -\frac{3v-1}{8} & -\frac{b^2v-b^2+4a^2}{12ab} & \frac{v+1}{8} & \frac{b^2v-b^2-2a^2}{12ab} \\
\frac{a^2v-2b^2-a^2}{12ab} & -\frac{v+1}{8} & \frac{a^2v+b^2-a^2}{6ab} & -\frac{3v-1}{8} & -\frac{a^2v-2b^2-a^2}{6ab} & \frac{v+1}{8} & -\frac{a^2v+4b^2-a^2}{12ab} & \frac{3v-1}{8} \\
-\frac{v+1}{8} & \frac{b^2v-b^2-2a^2}{12ab} & \frac{3v-1}{8} & -\frac{b^2v-b^2+4a^2}{12ab} & \frac{v+1}{8} & -\frac{b^2v-b^2-2a^2}{6ab} & -\frac{3v-1}{8} & \frac{b^2v-b^2+a^2}{6ab} \\
\frac{a^2v+b^2-a^2}{6ab} & \frac{3v-1}{8} & \frac{a^2v-2b^2-a^2}{12ab} & \frac{v+1}{8} & -\frac{a^2v+4b^2-a^2}{12ab} & -\frac{3v-1}{8} & -\frac{a^2v-2b^2-a^2}{6ab} & -\frac{v+1}{8} \\
-\frac{3v-1}{8} & -\frac{b^2v-b^2+4a^2}{12ab} & \frac{v+1}{8} & \frac{b^2v-b^2-2a^2}{12ab} & \frac{3v-1}{8} & \frac{b^2v-b^2+a^2}{6ab} & -\frac{v+1}{8} & -\frac{b^2v-b^2-2a^2}{6ab}
\end{bmatrix}
. \quad (3.76)
$$

3.2 Example: Two-dimensional rectangular element under plane strain condition

Reconsider Example 3.1 and derive the general expression for the elemental stiffness matrix under the plane strain assumption. Use only a 2×2 Gauss-Legendre integration rule.

3.2 Solution

The following Module 3.3 shows the entire wxMaxima code for the derivation of $\boldsymbol{K}^{\mathrm{e}}$ based on a 2×2 Gauss-Legendre integration rule

```
-->     kill(all);
        load("../Library/my_funs.mac")$
        ratprint : false$
(% o0)  done
```

Definition of Plane Stress Stiffness: Nodal Coordinates "ncoor", Material Properties "mat", Number of Element Nodes "nen", Thickness Function "thickness", Number of Quadrature Points "nqp"

```
-->     ncoor : [[-a,-b],[a,-b],[a,b],[-a,b]]$

        mat : [E, v]$

        thickness(x,y) := t$

        type : "plane strain"$
        nen : 4$
        nqp : 4$
```

Maxima Calculations and Output

```
-->     Ke :
        k_el_2D_plane_numerical(ncoor, mat, nqp, nen, type)$

        print(" ")$
        print("Element stiffness matrix:")$
        coeff : (E*t)/((1+v)*(1-2*v))$
        print(" ",coeff,factor(ratsimp(Ke/coeff)))$
```

Element Stiffness Matrix:

$$\frac{Et}{(1-2v)\,(v+1)}$$

$$\begin{bmatrix}
-\frac{2b^2v+2a^2v-2b^2-a^2}{6ab} & \frac{1}{2^3} & \frac{4b^2v-2a^2v-4b^2+a^2}{12ab} & \frac{4v-1}{8} & \cdots \\
\frac{1}{2^3} & -\frac{2b^2v+2a^2v-b^2-2a^2}{6ab} & -\frac{4v-1}{8} & \frac{2b^2v-a^2v-b^2+a^2}{6ab} & \cdots \\
\frac{4b^2v-2a^2v-4b^2+a^2}{12ab} & -\frac{4v-1}{8} & -\frac{2b^2v+2a^2v-2b^2-a^2}{6ab} & -\frac{1}{2^3} & \cdots \\
\frac{4v-1}{8} & \frac{2b^2v-a^2v-b^2+a^2}{6ab} & -\frac{1}{2^3} & -\frac{2b^2v+2a^2v-b^2-2a^2}{6ab} & \cdots \\
\frac{2b^2v+2a^2v-2b^2-a^2}{12ab} & -\frac{1}{2^3} & -\frac{b^2v-2a^2v-b^2+a^2}{6ab} & -\frac{4v-1}{8} & \cdots \\
-\frac{1}{2^3} & \frac{2b^2v+2a^2v-b^2-2a^2}{12ab} & \frac{4v-1}{8} & -\frac{2b^2v-4a^2v-b^2+4a^2}{12ab} & \cdots \\
-\frac{b^2v-2a^2v-b^2+a^2}{6ab} & \frac{4v-1}{8} & \frac{2b^2v+2a^2v-2b^2-a^2}{12ab} & \frac{1}{2^3} & \cdots \\
-\frac{4v-1}{8} & -\frac{2b^2v-4a^2v-b^2+4a^2}{12ab} & \frac{1}{2^3} & \frac{2b^2v+2a^2v-b^2-2a^2}{12ab} & \cdots
\end{bmatrix}$$

Module 3.3: Stiffness matrix for a two-dimensional rectangular element ($2a \times 2b$) under plane stress condition (2×2 numerical integration). The complete stiffness matrix is shown in Eq. (3.77)

$$
\boldsymbol{K}^{\mathrm{e}} = \frac{Et}{(1-2v)\,(v+1)} \times
$$

$$
\begin{bmatrix}
-\frac{2b^2v+2a^2v-2b^2-a^2}{6ab} & \frac{1}{2^3} & \frac{4b^2v-2a^2v-4b^2+a^2}{12ab} & \frac{4v-1}{8} & \frac{2b^2v+2a^2v-2b^2-a^2}{12ab} & -\frac{1}{2^3} & -\frac{b^2v-2a^2v-b^2+a^2}{6ab} & -\frac{4v-1}{8} \\
\frac{1}{2^3} & -\frac{2b^2v+2a^2v-b^2-2a^2}{6ab} & -\frac{4v-1}{8} & \frac{2b^2v-a^2v-b^2+a^2}{6ab} & -\frac{1}{2^3} & \frac{2b^2v+2a^2v-b^2-2a^2}{12ab} & \frac{4v-1}{8} & -\frac{2b^2v-4a^2v-b^2+4a^2}{12ab} \\
\frac{4b^2v-2a^2v-4b^2+a^2}{12ab} & -\frac{4v-1}{8} & -\frac{2b^2v+2a^2v-2b^2-a^2}{6ab} & -\frac{1}{2^3} & -\frac{b^2v-2a^2v-b^2+a^2}{6ab} & \frac{4v-1}{8} & \frac{2b^2v+2a^2v-2b^2-a^2}{12ab} & \frac{1}{2^3} \\
\frac{4v-1}{8} & \frac{2b^2v-a^2v-b^2+a^2}{6ab} & -\frac{1}{2^3} & -\frac{2b^2v+2a^2v-b^2-2a^2}{6ab} & -\frac{4v-1}{8} & -\frac{2b^2v-4a^2v-b^2+4a^2}{12ab} & \frac{1}{2^3} & \frac{2b^2v+2a^2v-b^2-2a^2}{12ab} \\
\frac{2b^2v+2a^2v-2b^2-a^2}{12ab} & -\frac{1}{2^3} & -\frac{b^2v-2a^2v-b^2+a^2}{6ab} & -\frac{4v-1}{8} & -\frac{2b^2v+2a^2v-2b^2-a^2}{6ab} & \frac{1}{2^3} & \frac{4b^2v-2a^2v-4b^2+a^2}{12ab} & \frac{4v-1}{8} \\
-\frac{1}{2^3} & \frac{2b^2v+2a^2v-b^2-2a^2}{12ab} & \frac{4v-1}{8} & -\frac{2b^2v-4a^2v-b^2+4a^2}{12ab} & \frac{1}{2^3} & -\frac{2b^2v+2a^2v-b^2-2a^2}{6ab} & -\frac{4v-1}{8} & \frac{2b^2v-a^2v-b^2+a^2}{6ab} \\
-\frac{b^2v-2a^2v-b^2+a^2}{6ab} & \frac{4v-1}{8} & \frac{2b^2v+2a^2v-2b^2-a^2}{12ab} & \frac{1}{2^3} & \frac{4b^2v-2a^2v-4b^2+a^2}{12ab} & -\frac{4v-1}{8} & -\frac{2b^2v+2a^2v-2b^2-a^2}{6ab} & -\frac{1}{2^3} \\
-\frac{4v-1}{8} & -\frac{2b^2v-4a^2v-b^2+4a^2}{12ab} & \frac{1}{2^3} & \frac{2b^2v+2a^2v-b^2-2a^2}{12ab} & \frac{4v-1}{8} & \frac{2b^2v-a^2v-b^2+a^2}{6ab} & -\frac{1}{2^3} & -\frac{2b^2v+2a^2v-b^2-2a^2}{6ab}
\end{bmatrix}. \tag{3.77}
$$

3.3 Example: Distorted two-dimensional element

Given is a distorted two-dimensional element as shown in Fig. 3.10. The constant thickness has a value of $t = 0.1$ and the materials parameters are given as $E = 70000$ and $\nu = 0.3$. Derive the general expression for the elemental stiffness matrix under the plane stress assumption. Use analytical, 2×2 Gauss-Legendre, and 3×3 Gauss-Legendre integration. Compare the accuracy of the integration approach for a few selected elements of the stiffness matrix.

3.3 Solution

The analytical solution gives the following expression for the elemental stiffness matrix:

$$\boldsymbol{K}^{\mathrm{e}} =$$

$$\begin{bmatrix} 3372.54 & 752.096 & -2390.34 & 52.1695 & -1091.86 & -1048.74 & 109.657 & 244.477 \\ 752.096 & 3372.54 & 244.477 & 109.657 & -1048.74 & -1091.86 & 52.1695 & -2390.34 \\ -2390.34 & 244.477 & 4373.81 & -1785.33 & 565.794 & -7.15985 & -2549.26 & 1548.01 \\ 52.1695 & 109.657 & -1785.33 & 4373.81 & 185.148 & -1934.21 & 1548.01 & -2549.26 \\ -1091.86 & -1048.74 & 565.794 & 185.148 & 2460.27 & 870.755 & -1934.21 & -7.15985 \\ -1048.74 & -1091.86 & -7.15985 & -1934.21 & 870.755 & 2460.27 & 185.148 & 565.794 \\ 109.657 & 52.1695 & -2549.26 & 1548.01 & -1934.21 & 185.148 & 4373.81 & -1785.33 \\ 244.477 & -2390.34 & 1548.01 & -2549.26 & -7.15985 & 565.794 & -1785.33 & 4373.81 \end{bmatrix} . \quad (3.78)$$

The following Module 3.4 shows the entire wxMaxima code for the derivation of $\boldsymbol{K}^{\mathrm{e}}$ based on a 2×2 Gauss-Legendre integration rule.
The following Module 3.5 shows the entire wxMaxima code for the derivation of $\boldsymbol{K}^{\mathrm{e}}$ based on a 3×3 Gauss-Legendre integration rule (see Table 3.3 for a comparison).

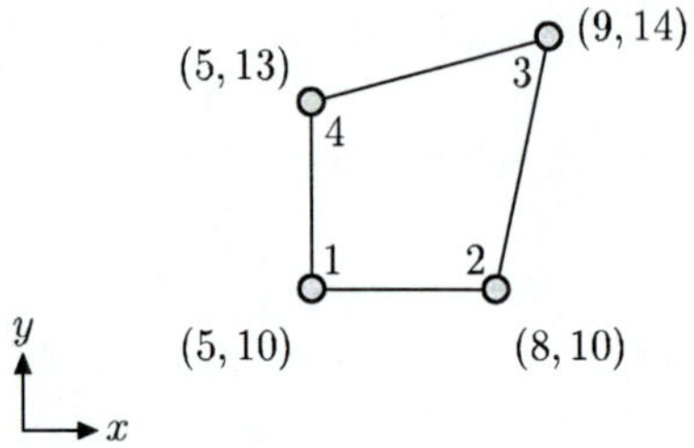

Fig. 3.10 Distorted two-dimensional element

Table 3.3 Influence of the integration on the accuracy

Element	Analytical	2×2 Numerical	Rel. Error in %	3×3 Numerical	Rel. Error in %
K_{11}	3372.54	3366.41	0.182	3372.52	5.930×10^{-4}
K_{12}	752.096	752.807	0.095	752.099	3.989×10^{-4}

```
→   kill(all);
    load("../Library/my_funs.mac")$
    fpprintprec : 6$ /* float numbers precision (only for printing) */
    ratprint : false$

(% o0) done
```

Definition of Plane Stress Stiffness: Nodal Coordinates "ncoor", Material Properties "mat", Number of Element Nodes "nen", Thickness Function "thickness", Number of Quadrature Points "nqp"

```
→   ncoor : [[5,10],[8,10],[9,14],[5,13]]$

    mat : [70000, 0.3]$

    thickness(x,y) := 0.1$

    type : "plane stress"$
    nen : 4$
    nqp : 4$
```

Maxima Calculations and Output

```
→   Ke :
    k_el_2D_plane_numerical(ncoor, mat, nqp, nen, type)$

    print(" ")$
    print("Element stiffness matrix:")$
    print(" ",float(factor(ratsimp(Ke))))$
```

Element Stiffness Matrix:

$$\begin{bmatrix} 3366.41 & 752.807 & -2385.43 & 51.6003 & -1095.54 & -1048.32 & 114.566 & 243.908 \\ 752.807 & 3366.41 & 243.908 & 114.566 & -1048.32 & -1095.54 & 51.6003 & -2385.43 \\ -2385.43 & 243.908 & 4369.89 & -1784.87 & 568.74 & -7.50136 & -2553.19 & 1548.46 \\ 51.6003 & 114.566 & -1784.87 & 4369.89 & 184.806 & -1931.26 & 1548.46 & -2553.19 \\ -1095.54 & -1048.32 & 568.74 & 184.806 & 2458.06 & 871.011 & -1931.26 & -7.50136 \\ -1048.32 & -1095.54 & -7.50136 & -1931.26 & 871.011 & 2458.06 & 184.806 & 568.74 \\ 114.566 & 51.6003 & -2553.19 & 1548.46 & -1931.26 & 184.806 & 4369.89 & -1784.87 \\ 243.908 & -2385.43 & 1548.46 & -2553.19 & -7.50136 & 568.74 & -1784.87 & 4369.89 \end{bmatrix}$$

Module 3.4: Stiffness matrix for a distorted two-dimensional element under plane stress condition (2×2 numerical integration)

```
→        kill(all);
         load("../Library/my_funs.mac")$
         fpprintprec : 6$ /* float numbers precision (only for printing) */
         ratprint : false$

(% o0)   done
```

Definition of Plane Stress Stiffness: Nodal Coordinates "ncoor", Material Properties "mat", Number of Element Nodes "nen", Thickness Function "thickness", Number of Quadrature Points "nqp"

```
→        ncoor : [[5,10],[8,10],[9,14],[5,13]]$

         mat : [70000, 0.3]$

         thickness(x,y) := 0.1$

         type : "plane stress"$
         nen : 4$
         nqp : 9$
```

Maxima Calculations and Output

```
→        Ke :
         k_el_2D_plane_numerical(ncoor, mat, nqp, nen, type)$

         print(" ")$
         print("Element stiffness matrix:")$
         print(" ",float(factor(ratsimp(Ke))))$
```

Element Stiffness Matrix:

$$\begin{bmatrix} 3372.52 & 752.099 & -2390.32 & 52.1673 & -1091.87 & -1048.74 & 109.677 & 244.475 \\ 752.099 & 3372.52 & 244.475 & 109.677 & -1048.74 & -1091.87 & 52.1673 & -2390.32 \\ -2390.32 & 244.475 & 4373.8 & -1785.32 & 565.806 & -7.16117 & -2549.28 & 1548.01 \\ 52.1673 & 109.677 & -1785.32 & 4373.8 & 185.147 & -1934.19 & 1548.01 & -2549.28 \\ -1091.87 & -1048.74 & 565.806 & 185.147 & 2460.26 & 870.755 & -1934.19 & -7.16117 \\ -1048.74 & -1091.87 & -7.16117 & -1934.19 & 870.755 & 2460.26 & 185.147 & 565.806 \\ 109.677 & 52.1673 & -2549.28 & 1548.01 & -1934.19 & 185.147 & 4373.8 & -1785.32 \\ 244.475 & -2390.32 & 1548.01 & -2549.28 & -7.16117 & 565.806 & -1785.32 & 4373.8 \end{bmatrix}$$

Module 3.5: Stiffness matrix for a distorted two-dimensional element under plane stress condition (3×3 numerical integration)

3.4 Example: Variable thickness of a two-dimensional element—dimensions given as variables

Given is a regular two-dimensional element as shown in Fig. 3.11. The thickness t is *not* constant for this problem.

Under the assumption that the origin of the x-y coordinate system is located in the center of the element, the function of the thickness t can be expressed as (see Fig. 3.12 for a graphical representation)

$$t(x, y) = \frac{a}{10}\left(1 + \frac{1}{4}\left(1 + \frac{x}{a}\right)\left(1 + \frac{y}{b}\right)\right), \tag{3.79}$$

where the geometrical dimensions a and b refer to the half width and height, respectively (see Fig. 3.11).

Derive the general expression for the elemental stiffness matrix under the plane stress and plane strain assumptions. Use only analytical integration for this problem.

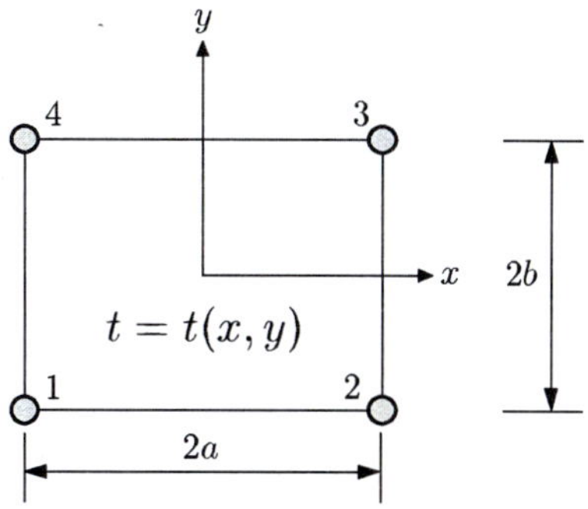

Fig. 3.11 Schematic representation of a rectangular element with variable thickness $t(x, y)$

3.4 Solution

The following Module 3.6 shows the entire wxMaxima code for the derivation of $\boldsymbol{K}^{\mathrm{e}}$ for the plane stress case based on analytical integration.

The following Module 3.7 shows the entire wxMaxima code for the derivation of $\boldsymbol{K}^{\mathrm{e}}$ for the plane strain case based on analytical integration.

```
→    kill(all);
     load("../Library/my_funs.mac")$
     ratprint : false$

(% o0)  done
```

Definition of Plane Stress Stiffness: Nodal Coordinates "ncoor", Material Properties "mat", Number of Element Nodes "nen", Thickness Function "thickness"

```
→    ncoor : [[-a,-b],[a,-b],[a,b],[-a,b]]$

     mat : [E, v]$

     thickness(x,y) := (a/10)*(1+(1/4)*(1+x/a)*(1+y/b))$

     type : "plane stress"$
     nen : 4$
```

Maxima Calculations and Output

```
→    Ke :
     k_el_2D_plane_analytical(ncoor, mat, nen, type)$

     print(" ")$
     print("Element stiffness matrix:")$
     coeff : E/(1-v^2)$
     print(" ",coeff,factor(ratsimp(Ke/coeff)))$
```

Element Stiffness Matrix:

$$\frac{E}{1-v^2}$$

$$\begin{bmatrix}
-\frac{3\left(a^2v-2b^2-a^2\right)}{160b} & \frac{a\,(v+1)}{72} & -\frac{5a^2v+18b^2-5a^2}{480b} & \frac{a\,(16v-5)}{360} & \cdots \\
\frac{a\,(v+1)}{72} & -\frac{3\left(b^2v-b^2-2a^2\right)}{160b} & -\frac{a\,(31v-11)}{720} & \frac{9b^2v-9b^2+10a^2}{480b} & \cdots \\
-\frac{5a^2v+18b^2-5a^2}{480b} & -\frac{a\,(31v-11)}{720} & -\frac{11a^2v-18b^2-11a^2}{480b} & -\frac{11a\,(v+1)}{720} & \cdots \\
\frac{a\,(16v-5)}{360} & \frac{9b^2v-9b^2+10a^2}{480b} & -\frac{11a\,(v+1)}{720} & -\frac{9b^2v-9b^2-22a^2}{480b} & \cdots \\
\frac{a^2v-2b^2-a^2}{96b} & -\frac{11a\,(v+1)}{720} & \frac{11a^2v+10b^2-11a^2}{480b} & -\frac{a\,(37v-11)}{720} & \cdots \\
-\frac{11a\,(v+1)}{720} & \frac{b^2v-b^2-2a^2}{96b} & \frac{a\,(35v-13)}{720} & -\frac{5b^2v-5b^2+22a^2}{480b} & \cdots \\
\frac{9a^2v+10b^2-9a^2}{480b} & \frac{a\,(16v-5)}{360} & \frac{a^2v-2b^2-a^2}{96b} & \frac{a\,(8v+5)}{360} & \cdots \\
-\frac{a\,(31v-11)}{720} & -\frac{5b^2v-5b^2+18a^2}{480b} & \frac{a\,(7v+13)}{720} & \frac{b^2v-b^2-2a^2}{96b} & \cdots
\end{bmatrix}$$

Module 3.6: Stiffness matrix for a two-dimensional element with variable thickness under plane stress condition (analytical integration). The complete stiffness matrix is shown in Eq. (3.80)

```
→       kill(all);
        load("../Library/my_funs.mac")$
        ratprint : false$

(% o0)  done
```

Definition of Plane Stress Stiffness: Nodal Coordinates "ncoor", Material Properties "mat", Number of Element Nodes "nen", Thickness Function "thickness"

```
→       ncoor : [[-a,-b],[a,-b],[a,b],[-a,b]]$

        mat : [E, v]$

        thickness(x,y) := (a/10)*(1+(1/4)*(1+x/a)*(1+y/b))$

        type : "plane strain"$
        nen : 4$
```

Maxima Calculations and Output

```
→       Ke :
        k_el_2D_plane_analytical(ncoor, mat, nen, type)$

        print(" ")$
        print("Element stiffness matrix:")$
        coeff : (E)/((1+v)*(1-2*v))$
        print(" ",coeff,factor(ratsimp(Ke/coeff)))$
```

Element Stiffness Matrix:

$$\frac{E}{(1-2v)\,(v+1)}$$

$$\begin{bmatrix}
-\frac{3\left(2b^2v+2a^2v-2b^2-a^2\right)}{160b} & \frac{a}{72} & \frac{18b^2v-10a^2v-18b^2+5a^2}{480b} & \cdots \\
\frac{a}{72} & -\frac{3\left(2b^2v+2a^2v-b^2-2a^2\right)}{160b} & -\frac{a\,(42v-11)}{720} & \cdots \\
\frac{18b^2v-10a^2v-18b^2+5a^2}{480b} & -\frac{a\,(42v-11)}{720} & -\frac{18b^2v+22a^2v-18b^2-11a^2}{480b} & \cdots \\
\frac{a\,(21v-5)}{360} & \frac{18b^2v-10a^2v-9b^2+10a^2}{480b} & -\frac{11a}{720} & \cdots \\
\frac{2b^2v+2a^2v-2b^2-a^2}{96b} & -\frac{11a}{720} & -\frac{10b^2v-22a^2v-10b^2+11a^2}{480b} & \cdots \\
-\frac{11a}{720} & \frac{2b^2v+2a^2v-b^2-2a^2}{96b} & \frac{a\,(48v-13)}{720} & \cdots \\
-\frac{10b^2v-18a^2v-10b^2+9a^2}{480b} & \frac{a\,(21v-5)}{360} & \frac{2b^2v+2a^2v-2b^2-a^2}{96b} & \cdots \\
-\frac{a\,(42v-11)}{720} & -\frac{10b^2v-18a^2v-5b^2+18a^2}{480b} & -\frac{a\,(6v-13)}{720} & \cdots
\end{bmatrix}$$

Module 3.7: Stiffness matrix for a two-dimensional element with variable thickness under plane strain condition (analytical integration). The complete stiffness matrix is shown in Eq. (3.81)

The complete stiffness matrix for the plane stress case based on analytical integration is shown in Eq. (3.80) while the plane strain case based on analytical integration is shown in Eq. (3.81).

$$
\boldsymbol{K}^{\mathrm{e}} = \frac{E}{1-v^2} \times
\begin{bmatrix}
-\frac{3\left(a^2v-2b^2-a^2\right)}{160b} & \frac{a\,(v+1)}{72} & -\frac{5a^2v+18b^2-5a^2}{480b} & \frac{a\,(16v-5)}{360} & \frac{a^2v-2b^2-a^2}{96b} & -\frac{11a\,(v+1)}{720} & \frac{9a^2v+10b^2-9a^2}{480b} & -\frac{a\,(31v-11)}{720} \\
\frac{a\,(v+1)}{72} & -\frac{3\left(b^2v-b^2-2a^2\right)}{160b} & -\frac{a\,(31v-11)}{720} & \frac{9b^2v-9b^2+10a^2}{480b} & -\frac{11a\,(v+1)}{720} & \frac{b^2v-b^2-2a^2}{96b} & \frac{a\,(16v-5)}{360} & -\frac{5b^2v-5b^2+18a^2}{480b} \\
-\frac{5a^2v+18b^2-5a^2}{480b} & -\frac{a\,(31v-11)}{720} & -\frac{11a^2v-18b^2-11a^2}{480b} & -\frac{11a\,(v+1)}{720} & \frac{11a^2v+10b^2-11a^2}{480b} & \frac{a\,(35v-13)}{720} & \frac{a^2v-2b^2-a^2}{96b} & \frac{a\,(7v+13)}{720} \\
\frac{a\,(16v-5)}{360} & \frac{9b^2v-9b^2+10a^2}{480b} & -\frac{11a\,(v+1)}{720} & -\frac{9b^2v-9b^2-22a^2}{480b} & -\frac{a\,(37v-11)}{720} & -\frac{5b^2v-5b^2+22a^2}{480b} & \frac{a\,(8v+5)}{360} & \frac{b^2v-b^2-2a^2}{96b} \\
\frac{a^2v-2b^2-a^2}{96b} & -\frac{11a\,(v+1)}{720} & \frac{11a^2v+10b^2-11a^2}{480b} & -\frac{a\,(37v-11)}{720} & -\frac{11\left(a^2v-2b^2-a^2\right)}{480b} & \frac{13a\,(v+1)}{720} & -\frac{5a^2v+22b^2-5a^2}{480b} & \frac{a\,(35v-13)}{720} \\
-\frac{11a\,(v+1)}{720} & \frac{b^2v-b^2-2a^2}{96b} & \frac{a\,(35v-13)}{720} & -\frac{5b^2v-5b^2+22a^2}{480b} & \frac{13a\,(v+1)}{720} & -\frac{11\left(b^2v-b^2-2a^2\right)}{480b} & -\frac{a\,(37v-11)}{720} & \frac{11b^2v-11b^2+10a^2}{480b} \\
\frac{9a^2v+10b^2-9a^2}{480b} & \frac{a\,(16v-5)}{360} & \frac{a^2v-2b^2-a^2}{96b} & \frac{a\,(8v+5)}{360} & -\frac{5a^2v+22b^2-5a^2}{480b} & -\frac{a\,(37v-11)}{720} & -\frac{9a^2v-22b^2-9a^2}{480b} & -\frac{11a\,(v+1)}{720} \\
-\frac{a\,(31v-11)}{720} & -\frac{5b^2v-5b^2+18a^2}{480b} & \frac{a\,(7v+13)}{720} & \frac{b^2v-b^2-2a^2}{96b} & \frac{a\,(35v-13)}{720} & \frac{11b^2v-11b^2+10a^2}{480b} & -\frac{11a\,(v+1)}{720} & -\frac{11b^2v-11b^2-18a^2}{480b}
\end{bmatrix}
\tag{3.80}
$$

$$
\boldsymbol{K}^{\mathrm{e}} = \frac{E}{(1-2v)\,(v+1)} \times
\begin{bmatrix}
-\frac{3(2b^2v+2a^2v-2b^2-a^2)}{160b} & \frac{a}{72} & \frac{18b^2v-10a^2v-18b^2+5a^2}{480b} & \frac{a\,(21v-5)}{360} & \frac{2b^2v+2a^2v-2b^2-a^2}{96b} & -\frac{11a}{720} & -\frac{10b^2v-18a^2v-10b^2+9a^2}{480b} & -\frac{a\,(42v-11)}{720} \\
\frac{a}{72} & -\frac{3(2b^2v+2a^2v-b^2-2a^2)}{160b} & -\frac{a\,(42v-11)}{720} & \frac{18b^2v-10a^2v-9b^2+10a^2}{480b} & -\frac{11a}{720} & \frac{2b^2v+2a^2v-b^2-2a^2}{96b} & \frac{a\,(21v-5)}{360} & -\frac{10b^2v-18a^2v-5b^2+18a^2}{480b} \\
\frac{18b^2v-10a^2v-18b^2+5a^2}{480b} & -\frac{a\,(42v-11)}{720} & -\frac{18b^2v+22a^2v-18b^2-11a^2}{480b} & -\frac{11a}{720} & -\frac{10b^2v-22a^2v-10b^2+11a^2}{480b} & \frac{a\,(48v-13)}{720} & \frac{2b^2v+2a^2v-2b^2-a^2}{96b} & -\frac{a\,(6v-13)}{720} \\
\frac{a\,(21v-5)}{360} & \frac{18b^2v-10a^2v-9b^2+10a^2}{480b} & -\frac{11a}{720} & -\frac{18b^2v+22a^2v-9b^2-22a^2}{480b} & -\frac{a\,(48v-11)}{720} & -\frac{10b^2v-22a^2v-5b^2+22a^2}{480b} & \frac{a\,(3v+5)}{360} & \frac{2b^2v+2a^2v-b^2-2a^2}{96b} \\
\frac{2b^2v+2a^2v-2b^2-a^2}{96b} & -\frac{11a}{720} & -\frac{10b^2v-22a^2v-10b^2+11a^2}{480b} & -\frac{a\,(48v-11)}{720} & -\frac{11(2b^2v+2a^2v-2b^2-a^2)}{480b} & \frac{13a}{720} & \frac{22b^2v-10a^2v-22b^2+5a^2}{480b} & \frac{a\,(48v-13)}{720} \\
-\frac{11a}{720} & \frac{2b^2v+2a^2v-b^2-2a^2}{96b} & \frac{a\,(48v-13)}{720} & -\frac{10b^2v-22a^2v-5b^2+22a^2}{480b} & \frac{13a}{720} & -\frac{11(2b^2v+2a^2v-b^2-2a^2)}{480b} & -\frac{a\,(48v-11)}{720} & \frac{22b^2v-10a^2v-11b^2+10a^2}{480b} \\
-\frac{10b^2v-18a^2v-10b^2+9a^2}{480b} & \frac{a\,(21v-5)}{360} & \frac{2b^2v+2a^2v-2b^2-a^2}{96b} & \frac{a\,(3v+5)}{360} & \frac{22b^2v-10a^2v-22b^2+5a^2}{480b} & -\frac{a\,(48v-11)}{720} & -\frac{22b^2v+18a^2v-22b^2-9a^2}{480b} & -\frac{11a}{720} \\
-\frac{a\,(42v-11)}{720} & -\frac{10b^2v-18a^2v-5b^2+18a^2}{480b} & -\frac{a\,(6v-13)}{720} & \frac{2b^2v+2a^2v-b^2-2a^2}{96b} & \frac{a\,(48v-13)}{720} & \frac{22b^2v-10a^2v-11b^2+10a^2}{480b} & -\frac{11a}{720} & -\frac{22b^2v+18a^2v-11b^2-18a^2}{480b}
\end{bmatrix}
\tag{3.81}
$$

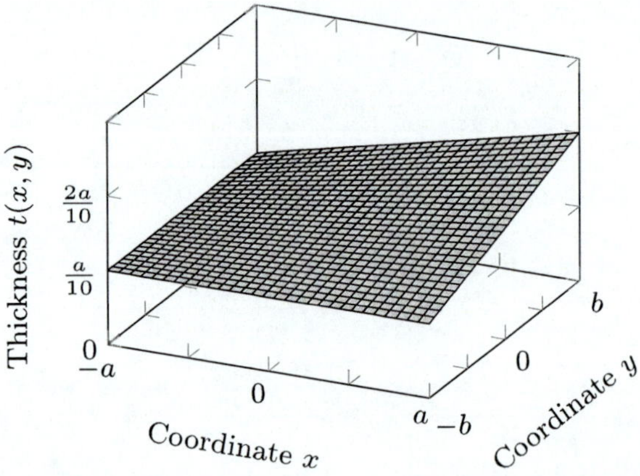

Fig. 3.12 Variable thickness of a two-dimensional element

3.5 Example: Variable thickness of a two-dimensional element—dimensions given as numbers

Given is again the regular two-dimensional element with variable thickness from Example 3.4. Consider the following numerical dimensions and properties: $a = 15$, $b = 10$, $E = 70000$, and $\nu = 0.3$. Derive the general expression for the elemental stiffness matrix under the plane stress assumption. Use analytical and 3×3 Gauss-Legendre integration. Compare the accuracy of the integration approach for a few selected elements of the stiffness matrix.

3.5 Solution

The following Module 3.8 shows the entire wxMaxima code for the derivation of $\boldsymbol{K}^{\mathrm{e}}$ for the plane stress case based on analytical integration and the given numerical values.

The following Module 3.9 shows the entire wxMaxima code for the derivation of $\boldsymbol{K}^{\mathrm{e}}$ for the plane stress case based on numerical 3×3 integration and the given numerical values.

```
→  kill(all);
   load("../Library/my_funs.mac")$
   fpprintprec : 6$ /* float numbers precision (only for printing) */
   ratprint : false$

(% o0)  done
```

Definition of Plane Stress Stiffness: Nodal Coordinates "ncoor", Material Properties "mat", Number of Element Nodes "nen", Thickness Function "thickness"

```
→  a : 15$
   b : 10$
   ncoor : [[-a,-b],[a,-b],[a,b],[-a,b]]$

   mat : [70000, 0.3]$

   thickness(x,y) := (a/10)*(1+(1/4)*(1+x/a)*(1+y/b))$

   type : "plane stress"$
   nen : 4$
```

Maxima Calculations and Output

```
→  Ke :
   k_el_2D_plane_analytical(ncoor, mat, nen, type)$

   print(" ")$
   print("Element stiffness matrix:")$
   print(" ", float(Ke))$
```

Element Stiffness Matrix:

$$\begin{bmatrix} 51562.5 & 20833.3 & -16226.0 & -641.026 & -28645.8 & -22916.7 & -6690.71 & 2724.36 \\ 20833.3 & 75000.0 & 2724.36 & 25961.5 & -22916.7 & -41666.7 & -641.026 & -59294.9 \\ -16226.0 & 2724.36 & 56610.6 & -22916.7 & -11738.8 & -4006.41 & -28645.8 & 24198.7 \\ -641.026 & 25961.5 & -22916.7 & 89423.1 & -160.256 & -73717.9 & 23717.9 & -41666.7 \\ -28645.8 & -22916.7 & -11738.8 & -160.256 & 63020.8 & 27083.3 & -22636.2 & -4006.41 \\ -22916.7 & -41666.7 & -4006.41 & -73717.9 & 27083.3 & 91666.7 & -160.256 & 23717.9 \\ -6690.71 & -641.026 & -28645.8 & 23717.9 & -22636.2 & -160.256 & 57972.8 & -22916.7 \\ 2724.36 & -59294.9 & 24198.7 & -41666.7 & -4006.41 & 23717.9 & -22916.7 & 77243.6 \end{bmatrix}$$

Module 3.8: Stiffness matrix for a two-dimensional element with variable thickness under plane stress condition (analytical integration and numerical values)

```
→   kill(all);
    load("../Library/my_funs.mac")$
    fpprintprec : 6$ /* float numbers precision (only for printing) */
    ratprint : false$

(% o0)  done
```

Definition of Plane Stress Stiffness: Nodal Coordinates "ncoor", Material Properties "mat", Number of Element Nodes "nen", Thickness Function "thickness", Number of Quadrature Points "nqp"

```
→   a : 15$
    b : 10$
    ncoor : [[-a,-b],[a,-b],[a,b],[-a,b]]$

    mat : [70000, 0.3]$

    thickness(x,y) := (a/10)*(1+(1/4)*(1+x/a)*(1+y/b))$

    type : "plane stress"$
    nen : 4$
    nqp : 9$
```

Maxima Calculations and Output

```
→   Ke :
    k_el_2D_plane_numerical(ncoor, mat, nqp, nen, type)$

    print(" ")$
    print("Element stiffness matrix:")$
    print(" ", float(Ke))$
```

Element Stiffness Matrix:

$$\begin{bmatrix} 51562.5 & 20833.3 & -16226.0 & -641.026 & -28645.8 & -22916.7 & -6690.71 & 2724.36 \\ 20833.3 & 75000.0 & 2724.36 & 25961.5 & -22916.7 & -41666.7 & -641.026 & -59294.9 \\ -16226.0 & 2724.36 & 56610.6 & -22916.7 & -11738.8 & -4006.41 & -28645.8 & 24198.7 \\ -641.026 & 25961.5 & -22916.7 & 89423.1 & -160.256 & -73717.9 & 23717.9 & -41666.7 \\ -28645.8 & -22916.7 & -11738.8 & -160.256 & 63020.8 & 27083.3 & -22636.2 & -4006.41 \\ -22916.7 & -41666.7 & -4006.41 & -73717.9 & 27083.3 & 91666.7 & -160.256 & 23717.9 \\ -6690.71 & -641.026 & -28645.8 & 23717.9 & -22636.2 & -160.256 & 57972.8 & -22916.7 \\ 2724.36 & -59294.9 & 24198.7 & -41666.7 & -4006.41 & 23717.9 & -22916.7 & 77243.6 \end{bmatrix}$$

Module 3.9: Stiffness matrix for a two-dimensional element with variable thickness under plane stress condition (numerical 3×3 integration and numerical values)

3.6 Example: Variable thickness of a two-dimensional element—dimensions given as numbers (version 2)

Given is a regular two-dimensional element as shown in Fig. 3.13. The thickness t is *not* constant for this problem.

Under the assumption that the origin of the x-y coordinate system is located in the center of the element, the function of the thickness t can be expressed as (see Fig. 3.14 for a graphical representation)

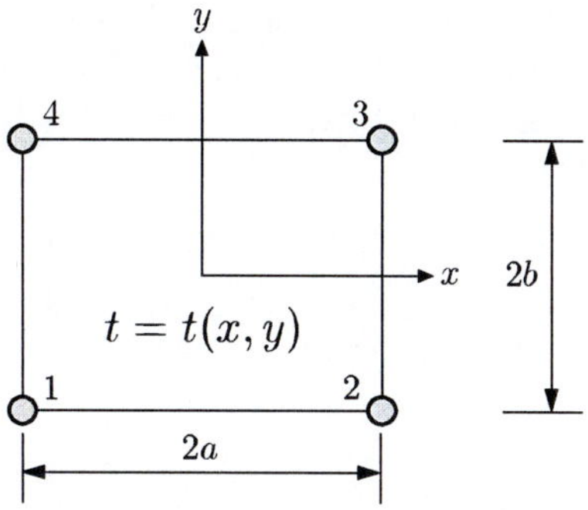

Fig. 3.13 Schematic representation of a rectangular element with variable thickness $t(x, y)$ (version 2)

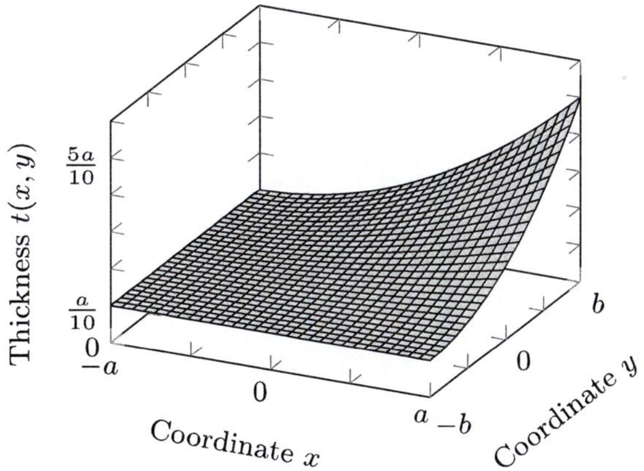

Fig. 3.14 Variable thickness of a two-dimensional element (version 2)

$$t(x, y) = \frac{a}{10}\left(1 + \frac{1}{4}\left(1 + \frac{x}{a}\right)^2\left(1 + \frac{y}{b}\right)^2\right), \tag{3.82}$$

where the geometrical dimensions a and b refer to the half width and height, respectively (see Fig. 3.13).

Consider the following numerical dimensions and properties: $a = 15$, $b = 10$, $E = 70000$, and $\nu = 0.3$. Derive the general expression for the elemental stiffness matrix under the plane stress assumption. Use analytical, 2×2 Gauss-Legendre, and 3×3 Gauss-Legendre integration. Compare the accuracy of the integration approach for a few selected elements of the stiffness matrix.

3.6 Solution

The following Module 3.10 shows the entire wxMaxima code for the derivation of $\boldsymbol{K}^{\mathrm{e}}$ for the plane stress case based on analytical integration and the given numerical values.

The following Module 3.11 shows the entire wxMaxima code for the derivation of $\boldsymbol{K}^{\mathrm{e}}$ for the plane stress case based on a numerical 2×2 integration and the given numerical values.

The following Module 3.12 shows the entire wxMaxima code for the derivation of $\boldsymbol{K}^{\mathrm{e}}$ for the plane stress case based on a numerical 3×3 integration and the given numerical values.

```
--> kill(all);
    load("../Library/my_funs.mac")$
    fpprintprec : 6$ /* float numbers precision (only for printing) */
    ratprint : false$

(% o0) done
```

Definition of Plane Stress Stiffness: Nodal Coordinates "ncoor", Material Properties "mat", Number of Element Nodes "nen", Thickness Function "thickness"

```
--> a : 15$
    b : 10$
    ncoor : [[-a,-b],[a,-b],[a,b],[-a,b]]$

    mat : [70000, 0.3]$

    thickness(x,y) := (a/10)*(1+(1/4)*(1+x/a)^2*(1+y/b)^2)$

    type : "plane stress"$
    nen : 4$
```

Maxima Calculations and Output

```
--> Ke :
    k_el_2D_plane_analytical(ncoor, mat, nen, type)$

    print(" ")$
    print("Element stiffness matrix:")$
    print(" ", float(Ke))$
```

Element Stiffness Matrix:

$$\begin{bmatrix} 51944.4 & 20833.3 & -14925.2 & 320.513 & -32083.3 & -25000.0 & -4935.9 & 3846.15 \\ 20833.3 & 75555.6 & 3846.15 & 30213.7 & -25000.0 & -46666.7 & 320.513 & -59102.6 \\ -14925.2 & 3846.15 & 65406.0 & -25000.0 & -18397.4 & -8653.85 & -32083.3 & 29807.7 \\ 320.513 & 30213.7 & -25000.0 & 1.14017\times 10^{5} & -3846.15 & -97564.1 & 28525.6 & -46666.7 \\ -32083.3 & -25000.0 & -18397.4 & -3846.15 & 82500.0 & 37500.0 & -32019.2 & -8653.85 \\ -25000.0 & -46666.7 & -8653.85 & -97564.1 & 37500.0 & 1.2\times 10^{5} & -3846.15 & 24230.8 \\ -4935.9 & 320.513 & -32083.3 & 28525.6 & -32019.2 & -3846.15 & 69038.5 & -25000.0 \\ 3846.15 & -59102.6 & 29807.7 & -46666.7 & -8653.85 & 24230.8 & -25000.0 & 81538.5 \end{bmatrix}$$

Module 3.10: Stiffness matrix for a two-dimensional element with variable thickness (version 2) under plane stress condition (analytical integration and numerical values)

```
-->   kill(all);
      load("../Library/my_funs.mac")$
      fpprintprec : 6$ /* float numbers precision (only for printing) */
      ratprint : false$

(% o0) done
```

Definition of Plane Stress Stiffness: Nodal Coordinates "ncoor", Material Properties "mat", Number of Element Nodes "nen", Thickness Function "thickness", Number of Quadrature Points "nqp"

```
-->   a : 15$
      b : 10$
      ncoor : [[-a,-b],[a,-b],[a,b],[-a,b]]$

      mat : [70000, 0.3]$

      thickness(x,y) := (a/10)*(1+(1/4)*(1+x/a)^2*(1+y/b)^2)$

      type : "plane stress"$
      nen : 4$
      nqp : 4$
```

Maxima Calculations and Output

```
-->   Ke :
      k_el_2D_plane_analytical(ncoor, mat, nen, type)$

      print(" ")$
      print("Element stiffness matrix:")$
      print(" ", float(Ke))$
```

Element Stiffness Matrix:

$$\begin{bmatrix} 50925.9 & 20833.3 & -13906.7 & 320.513 & -33101.9 & -25000.0 & -3917.38 & 3846.15 \\ 20833.3 & 74074.1 & 3846.15 & 31695.2 & -25000.0 & -48148.1 & 320.513 & -57621.1 \\ -13906.7 & 3846.15 & 64387.5 & -25000.0 & -17378.9 & -8653.85 & -33101.9 & 29807.7 \\ 320.513 & 31695.2 & -25000.0 & 1.12536\times10^5 & -3846.15 & -96082.6 & 28525.6 & -48148.1 \\ -33101.9 & -25000.0 & -17378.9 & -3846.15 & 81481.5 & 37500.0 & -31000.7 & -8653.85 \\ -25000.0 & -48148.1 & -8653.85 & -96082.6 & 37500.0 & 1.18519\times10^5 & -3846.15 & 25712.3 \\ -3917.38 & 320.513 & -33101.9 & 28525.6 & -31000.7 & -3846.15 & 68019.9 & -25000.0 \\ 3846.15 & -57621.1 & 29807.7 & -48148.1 & -8653.85 & 25712.3 & -25000.0 & 80057.0 \end{bmatrix}$$

Module 3.11: Stiffness matrix for a two-dimensional element with variable thickness (version 2) under plane stress condition (numerical 2×2 integration and numerical values)

```
→       kill(all);
        load("../Library/my_funs.mac")$
        fpprintprec : 6$ /* float numbers precision (only for printing) */
        ratprint : false$

(% o0)  done
```

Definition of Plane Stress Stiffness: Nodal Coordinates "ncoor", Material Properties "mat", Number of Element Nodes "nen", Thickness Function "thickness", Number of Quadrature Points "nqp"

```
→       a : 15$
        b : 10$
        ncoor : [[-a,-b],[a,-b],[a,b],[-a,b]]$

        mat : [70000, 0.3]$

        thickness(x,y) := (a/10)*(1+(1/4)*(1+x/a)^2*(1+y/b)^2)$

        type : "plane stress"$
        nen : 4$
        nqp : 9$
```

Maxima Calculations and Output

```
→       Ke :
        k_el_2D_plane_analytical(ncoor, mat, nen, type)$

        print(" ")$
        print("Element stiffness matrix:")$
        print(" ", float(Ke))$
```

Element Stiffness Matrix:

$$\begin{bmatrix} 51944.4 & 20833.3 & -14925.2 & 320.513 & -32083.3 & -25000.0 & -4935.9 & 3846.15 \\ 20833.3 & 75555.6 & 3846.15 & 30213.7 & -25000.0 & -46666.7 & 320.513 & -59102.6 \\ -14925.2 & 3846.15 & 65406.0 & -25000.0 & -18397.4 & -8653.85 & -32083.3 & 29807.7 \\ 320.513 & 30213.7 & -25000.0 & 1.14017\times 10^5 & -3846.15 & -97564.1 & 28525.6 & -46666.7 \\ -32083.3 & -25000.0 & -18397.4 & -3846.15 & 82500.0 & 37500.0 & -32019.2 & -8653.85 \\ -25000.0 & -46666.7 & -8653.85 & -97564.1 & 37500.0 & 1.2\times 10^5 & -3846.15 & 24230.8 \\ -4935.9 & 320.513 & -32083.3 & 28525.6 & -32019.2 & -3846.15 & 69038.5 & -25000.0 \\ 3846.15 & -59102.6 & 29807.7 & -46666.7 & -8653.85 & 24230.8 & -25000.0 & 81538.5 \end{bmatrix}$$

Module 3.12: Stiffness matrix for a two-dimensional element with variable thickness (version 2) under plane stress condition (numerical 3×3 integration and numerical values)

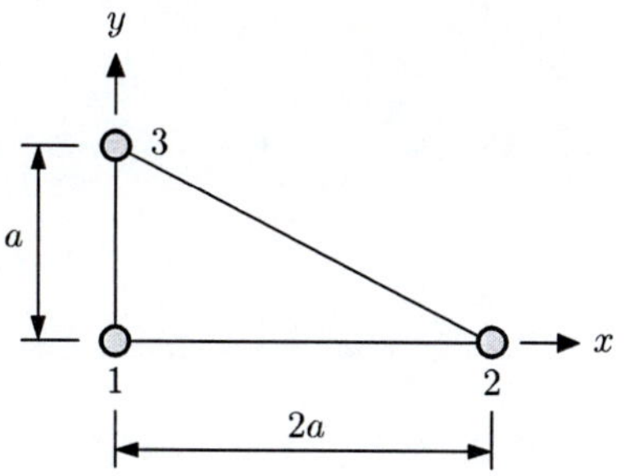

Fig. 3.15 Two-dimensional triangular element

3.7 Example: Two-dimensional triangular element under plane stress and plane strain conditions
Given is a regular two-dimensional element as shown in Fig. 3.15. The geometry can be described by the length parameter a.

Derive the general expression for the elemental stiffness matrix under the plane stress and plane strain assumptions. Furthermore, it can be assumed that the thickness t is constant. Use analytical integration to obtain the stiffness matrix.

3.7 Solution
The following Module 3.13 shows the entire wxMaxima code for the derivation of $\boldsymbol{K}^{\mathrm{e}}$ for the plane stress case based on analytical integration.
The following Module 3.14 shows the entire wxMaxima code for the derivation of $\boldsymbol{K}^{\mathrm{e}}$ for the plane strain case based on analytical integration.

```
-->  kill(all);
     load("../Library/my_funs.mac")$
     fpprintprec : 6$ /* float numbers precision (only for printing) */
     ratprint : false$
```

(% o0) *done*

Definition of Plane Stress Stiffness: Nodal Coordinates "ncoor", Material Properties "mat", Number of Element Nodes "nen", Thickness Function "thickness"

```
-->  ncoor : [[0,0],[2*a,0],[0,a]]$

     mat : [E, v]$

     thickness(x,y) := t$

     type : "plane stress"$
     nen : 3$
```

Maxima Calculations and Output

```
-->  Ke :
     k_el_2D_plane_analytical(ncoor, mat, nen, type)$

     print(" ")$
     print("Element Stiffness Matrix:")$
     print(" ", factor(Ke))$
```

Element Stiffness Matrix:

$$
\begin{bmatrix}
\frac{Et\,(2v-3)}{4(v-1)\,(v+1)} & -\frac{Et}{4(v-1)} & \frac{Et}{4(v-1)\,(v+1)} & -\frac{Et}{4(v+1)} & -\frac{Et}{2(v+1)} & \frac{Etv}{2(v-1)\,(v+1)} \\
-\frac{Et}{4(v-1)} & \frac{Et\,(v-9)}{8(v-1)\,(v+1)} & \frac{Etv}{2(v-1)\,(v+1)} & -\frac{Et}{8(v+1)} & -\frac{Et}{4(v+1)} & \frac{Et}{(v-1)\,(v+1)} \\
\frac{Et}{4(v-1)\,(v+1)} & \frac{Etv}{2(v-1)\,(v+1)} & -\frac{Et}{4(v-1)\,(v+1)} & 0 & 0 & -\frac{Etv}{2(v-1)\,(v+1)} \\
-\frac{Et}{4(v+1)} & -\frac{Et}{8(v+1)} & 0 & \frac{Et}{8(v+1)} & \frac{Et}{4(v+1)} & 0 \\
-\frac{Et}{2(v+1)} & -\frac{Et}{4(v+1)} & 0 & \frac{Et}{4(v+1)} & \frac{Et}{2(v+1)} & 0 \\
\frac{Etv}{2(v-1)\,(v+1)} & \frac{Et}{(v-1)\,(v+1)} & -\frac{Etv}{2(v-1)\,(v+1)} & 0 & 0 & -\frac{Et}{(v-1)\,(v+1)}
\end{bmatrix}
$$

Module 3.13: Stiffness matrix for a two-dimensional triangular element under the plane stress condition (analytical integration)

```
→   kill(all);
    load("../Library/my_funs.mac")$
    fpprintprec : 6$ /* float numbers precision (only for printing) */
    ratprint : false$

(%o0)  done
```

Definition of Plane Strain Stiffness: Nodal Coordinates "ncoor", Material Properties "mat", Number of Element Nodes "nen", Thickness Function "thickness"

```
→   ncoor : [[0,0],[2*a,0],[0,a]]$

    mat : [E, v]$

    thickness(x,y) := t$

    type : "plane strain"$
    nen : 3$
```

Maxima Calculations and Output

```
→   Ke :
    k_el_2D_plane_analytical(ncoor, mat, nen, type)$

    print(" ")$
    print("Element Stiffness Matrix:")$
    print(" ", factor(Ke))$
```

Element Stiffness Matrix:

$$\begin{bmatrix} \frac{Et\,(5v-3)}{4(v+1)\,(2v-1)} & -\frac{Et}{4(v+1)\,(2v-1)} & -\frac{Et\,(v-1)}{4(v+1)\,(2v-1)} & -\frac{Et}{4(v+1)} & -\frac{Et}{2(v+1)} & \frac{Etv}{2(v+1)\,(2v-1)} \\ -\frac{Et}{4(v+1)\,(2v-1)} & \frac{Et\,(10v-9)}{8(v+1)\,(2v-1)} & \frac{Etv}{2(v+1)\,(2v-1)} & -\frac{Et}{8(v+1)} & -\frac{Et}{4(v+1)} & -\frac{Et\,(v-1)}{(v+1)\,(2v-1)} \\ -\frac{Et\,(v-1)}{4(v+1)\,(2v-1)} & \frac{Etv}{2(v+1)\,(2v-1)} & \frac{Et\,(v-1)}{4(v+1)\,(2v-1)} & 0 & 0 & -\frac{Etv}{2(v+1)\,(2v-1)} \\ -\frac{Et}{4(v+1)} & -\frac{Et}{8(v+1)} & 0 & \frac{Et}{8(v+1)} & \frac{Et}{4(v+1)} & 0 \\ -\frac{Et}{2(v+1)} & -\frac{Et}{4(v+1)} & 0 & \frac{Et}{4(v+1)} & \frac{Et}{2(v+1)} & 0 \\ \frac{Etv}{2(v+1)\,(2v-1)} & -\frac{Et\,(v-1)}{(v+1)\,(2v-1)} & -\frac{Etv}{2(v+1)\,(2v-1)} & 0 & 0 & \frac{Et\,(v-1)}{(v+1)\,(2v-1)} \end{bmatrix}$$

Module 3.14: Stiffness matrix for a two-dimensional triangular element under the plane strain condition (analytical integration)

3.2 Elemental and Global Principal Finite Element Equation

3.2.1 Elements with Four Nodes

The principal finite element equation for a single linear four-node planar element can be stated as follows (see [5])

$$\begin{bmatrix} K_{11} & K_{12} & \dots & K_{17} & K_{18} \\ K_{21} & K_{22} & \dots & K_{27} & K_{28} \\ \vdots & \vdots & & \vdots & \vdots \\ K_{71} & K_{72} & \dots & K_{77} & K_{78} \\ K_{81} & K_{82} & \dots & K_{87} & K_{88} \end{bmatrix} \begin{bmatrix} u_{1x} \\ u_{1y} \\ \vdots \\ u_{4x} \\ u_{4y} \end{bmatrix} = \int_V \begin{bmatrix} N_1 & 0 \\ 0 & N_1 \\ \hline & \\ \vdots & \\ \hline N_4 & 0 \\ 0 & N_4 \end{bmatrix} \begin{bmatrix} f_x \\ f_y \end{bmatrix} \mathrm{d}V$$

$$+ \int_S \begin{bmatrix} N_1 & 0 \\ 0 & N_1 \\ \hline & \\ \vdots & \\ \hline N_4 & 0 \\ 0 & N_4 \end{bmatrix} \begin{bmatrix} t_x \\ t_y \end{bmatrix} \mathrm{d}S\,, \tag{3.83}$$

or in abbreviated form

$$\boldsymbol{K}^{\mathrm{e}} \boldsymbol{u}_{\mathrm{p}}^{\mathrm{e}} = \boldsymbol{f}_{\boldsymbol{b}}^{\mathrm{e}} + \boldsymbol{f}_{\boldsymbol{t}}^{\mathrm{e}}\,, \tag{3.84}$$

where $\boldsymbol{K}^{\mathrm{e}}$ is the elemental stiffness matrix, $\boldsymbol{u}_{\mathrm{p}}^{\mathrm{e}}$ is the column matrix of nodal unknowns, $\boldsymbol{f}_{\boldsymbol{b}}^{\mathrm{e}}$ is body force column matrix, and $\boldsymbol{f}_{\boldsymbol{t}}^{\mathrm{e}}$ is the boundary force matrix. The four interpolation functions N_i are given in Eqs. (3.12)–(3.15).

Let us have now a closer look on the boundary force matrix $\boldsymbol{f}_{\boldsymbol{t}}^{\mathrm{e}}$. The integral

$$\boldsymbol{f}_{\boldsymbol{t}}^{\mathrm{e}} = \int_S \boldsymbol{N}\boldsymbol{t}\,\mathrm{d}S\,, \tag{3.85}$$

which represents a surface integral, can be interpreted in different ways. One way is to understand it as single forces which are applied at nodes.[5] Then, we can write Eq. (3.85) in the following way (see Fig. 3.16a):

[5] See [5] for further details.

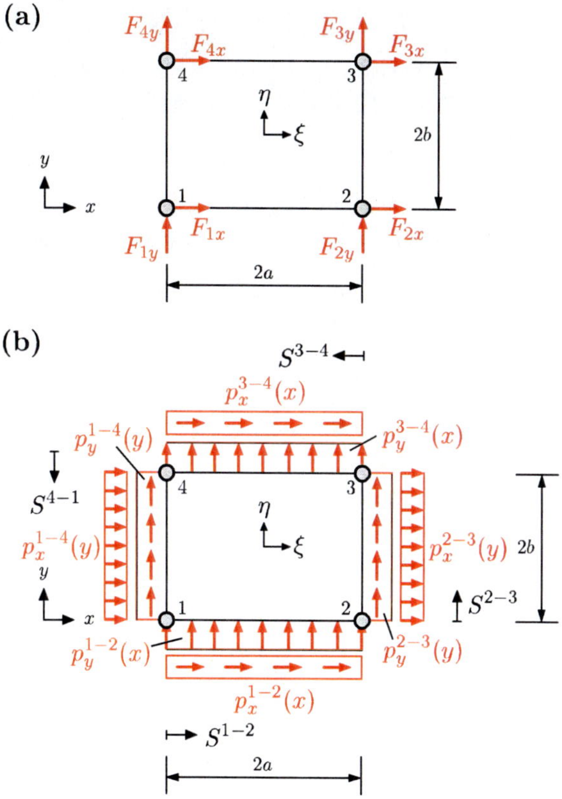

Fig. 3.16 Natural boundary conditions: **a** concentrated forces applied at single nodes, **b** distributed pressures at the edges of the element (each force and pressure is sketched in its positive direction)

$$
\boldsymbol{f}_t^{\mathrm{e}} = \int_S \boldsymbol{N}\boldsymbol{t}\,\mathrm{d}S = \begin{bmatrix} F_{1x} \\ F_{1y} \\ F_{2x} \\ F_{2y} \\ F_{3x} \\ F_{3y} \\ F_{4x} \\ F_{4y} \end{bmatrix}. \tag{3.86}
$$

Another view can be applied in the case of distributed pressures at the edges of the element, see Fig. 3.16b. Then we can write the surface integral in the following way:

$$
\int_S \boldsymbol{N}\boldsymbol{t}\,\mathrm{d}S = \int_{1-2} \boldsymbol{N}|_{\xi,\eta=-1} \begin{bmatrix} p_x^{1-2}(x) \\ p_y^{1-2}(x) \end{bmatrix} \mathrm{d}S^{1-2} + \int_{2-3} \boldsymbol{N}|_{\xi=1,\eta} \begin{bmatrix} p_x^{2-3}(y) \\ p_y^{2-3}(y) \end{bmatrix} \mathrm{d}S^{2-3}
$$

$$+\int\limits_{3-4} \boldsymbol{N}|_{\xi,\eta=1} \begin{bmatrix} p_x^{3-4}(x) \\ p_y^{3-4}(x) \end{bmatrix} \mathrm{d}S^{3-4} + \int\limits_{4-1} \boldsymbol{N}|_{\xi=-1,\eta} \begin{bmatrix} p_x^{4-1}(y) \\ p_y^{4-1}(y) \end{bmatrix} \mathrm{d}S^{4-1},$$

$$= \int\limits_{1-2} \begin{bmatrix} N_1 & 0 \\ 0 & N_1 \\ \hline N_2 & 0 \\ 0 & N_2 \\ \hline 0 & 0 \\ 0 & 0 \\ \hline 0 & 0 \\ 0 & 0 \end{bmatrix} \begin{bmatrix} p_x^{1-2}(x) \\ p_y^{1-2}(x) \end{bmatrix} \mathrm{d}S^{1-2} + \int\limits_{2-3} \begin{bmatrix} 0 & 0 \\ 0 & 0 \\ \hline N_2 & 0 \\ 0 & N_2 \\ \hline N_3 & 0 \\ 0 & N_3 \\ \hline 0 & 0 \\ 0 & 0 \end{bmatrix} \begin{bmatrix} p_x^{2-3}(y) \\ p_y^{2-3}(y) \end{bmatrix} \mathrm{d}S^{2-3} \tag{3.87}$$

$$+ \int\limits_{3-4} \begin{bmatrix} 0 & 0 \\ 0 & 0 \\ \hline 0 & 0 \\ 0 & 0 \\ \hline N_3 & 0 \\ 0 & N_3 \\ \hline N_4 & 0 \\ 0 & N_4 \end{bmatrix} \begin{bmatrix} p_x^{3-4}(x) \\ p_y^{3-4}(x) \end{bmatrix} \mathrm{d}S^{3-4} + \int\limits_{4-1} \begin{bmatrix} N_1 & 0 \\ 0 & N_1 \\ \hline 0 & 0 \\ 0 & 0 \\ \hline 0 & 0 \\ 0 & 0 \\ \hline N_4 & 0 \\ 0 & N_4 \end{bmatrix} \begin{bmatrix} p_x^{4-1}(y) \\ p_y^{4-1}(y) \end{bmatrix} \mathrm{d}S^{4-1},$$

or finally after the matrix multiplication:

$$\int\limits_S \boldsymbol{N}\boldsymbol{t}\,\mathrm{d}S = \int\limits_{1-2} \begin{bmatrix} N_1 p_x^{1-2}(x) \\ N_1 p_y^{1-2}(x) \\ \hline N_2 p_x^{1-2}(x) \\ N_2 p_y^{1-2}(x) \\ \hline 0 \\ 0 \\ \hline 0 \\ 0 \end{bmatrix}_{\xi,\eta=-1} \mathrm{d}S^{1-2} + \int\limits_{2-3} \begin{bmatrix} 0 \\ 0 \\ \hline N_2 p_x^{2-3}(y) \\ N_2 p_y^{2-3}(y) \\ \hline N_3 p_x^{2-3}(y) \\ N_3 p_y^{2-3}(y) \\ \hline 0 \\ 0 \end{bmatrix}_{\xi=1,\eta} \mathrm{d}S^{2-3}$$

$$
+\int\limits_{3-4}\begin{bmatrix} 0 \\ 0 \\ \hline 0 \\ 0 \\ \hline N_3 p_x^{3-4}(x) \\ N_3 p_y^{3-4}(x) \\ \hline N_4 p_x^{3-4}(x) \\ N_4 p_y^{3-4}(x) \end{bmatrix}_{\xi,\eta=1} \mathrm{d}S^{3-4} + \int\limits_{4-1}\begin{bmatrix} N_1 p_x^{4-1}(y) \\ N_1 p_y^{4-1}(y) \\ \hline 0 \\ 0 \\ \hline 0 \\ 0 \\ \hline N_4 p_x^{4-1}(y) \\ N_4 p_y^{4-1}(y) \end{bmatrix}_{\xi=-1,\eta} \mathrm{d}S^{4-1}\,. \tag{3.88}
$$

To transform the argument of the integration from $S = s \times t$ under the assumption of a constant thickness t to the natural coordinates ξ or η, we need to consider the relationships as indicated in Tables 3.4 and 3.5.

From Tables 3.4 and 3.5 we conclude the following relationships between natural and local coordinates:

$$
\xi = \frac{2s^{1-2}}{2a} - 1 \qquad \text{or} \qquad \mathrm{d}s^{1-2} = a\mathrm{d}\xi\,, \tag{3.89}
$$

$$
\eta = \frac{2s^{2-3}}{2b} - 1 \qquad \text{or} \qquad \mathrm{d}s^{2-3} = b\mathrm{d}\eta\,, \tag{3.90}
$$

$$
\xi = -\frac{2s^{3-4}}{2a} + 1 \qquad \text{or} \qquad \mathrm{d}s^{3-4} = -a\mathrm{d}\xi\,, \tag{3.91}
$$

$$
\eta = -\frac{2s^{4-1}}{2b} + 1 \qquad \text{or} \qquad \mathrm{d}s^{4-1} = -b\mathrm{d}\xi\,. \tag{3.92}
$$

Thus, under consideration of the unit space and the assumption that the thickness t is constant:

Table 3.4 Relationship between natural coordinate ξ and local coordinate s^{1-2} (see Fig. 3.16b)

Node	Coordinate ξ	Coordinate s^{1-2}
1	-1	0
2	$+1$	$2a$

Table 3.5 Relationship between natural coordinate η and local coordinate s^{4-1} (see Fig. 3.16b)

Node	Coordinate η	Coordinate s^{4-1}
4	$+1$	0
1	-1	$2b$

$$\int_S \boldsymbol{N}\boldsymbol{t}\,\mathrm{d}S = \int_{-1}^{+1} \left[\begin{array}{c} N_1(\xi)p_x^{1-2}(\xi) \\ N_1(\xi)p_y^{1-2}(\xi) \\ \hline N_2(\xi)p_x^{1-2}(\xi) \\ N_2(\xi)p_y^{1-2}(\xi) \\ \hline 0 \\ 0 \\ \hline 0 \\ 0 \end{array}\right]_{\xi,\eta=-1} t \underbrace{J}_{a} \mathrm{d}\xi + \int_{-1}^{+1} \left[\begin{array}{c} 0 \\ 0 \\ \hline N_2(\eta)p_x^{2-3}(\eta) \\ N_2(\eta)p_y^{2-3}(\eta) \\ \hline N_3(\eta)p_x^{2-3}(\eta) \\ N_3(\eta)p_y^{2-3}(v) \\ \hline 0 \\ 0 \end{array}\right]_{\xi=1,\eta} t \underbrace{J}_{b} \mathrm{d}\eta$$

$$+ \int_{+1}^{-1} \left[\begin{array}{c} 0 \\ 0 \\ \hline 0 \\ 0 \\ \hline N_3(\xi)p_x^{3-4}(\xi) \\ N_3(\xi)p_y^{3-4}(\xi) \\ \hline N_4(\xi)p_x^{3-4}(\xi) \\ N_4(\xi)p_y^{3-4}(\xi) \end{array}\right]_{\xi,\eta=1} t \underbrace{J}_{-a} \mathrm{d}\xi + \int_{+1}^{-1} \left[\begin{array}{c} N_1(\eta)p_x^{4-1}(\eta) \\ N_1(\eta)p_y^{4-1}(\eta) \\ \hline 0 \\ 0 \\ \hline 0 \\ 0 \\ \hline N_4(\eta)p_x^{4-1}(\eta) \\ N_4(\eta)p_y^{4-1}(\eta) \end{array}\right]_{\xi=-1,\eta} t \underbrace{J}_{-b} \mathrm{d}\eta\,. \tag{3.93}$$

It should be noted here that Eq. (3.93) is only valid for the special case of a regular four-node planar element at which the edges are parallel to the coordinate axes x and y (see Fig. 3.16). In the most general case, i.e., an arbitrary four-node planar element (see Fig. 3.17), Eq. (3.93) must be generalized by considering the appropriate Jacobians (see Fig. 3.17 for the definition of the edge lengths):

$$\int_S \boldsymbol{N}\boldsymbol{t}\,\mathrm{d}S = \int_{-1}^{+1} \left[\begin{array}{c} N_1(\xi)p_x^{1-2}(\xi) \\ N_1(\xi)p_y^{1-2}(\xi) \\ \hline N_2(\xi)p_x^{1-2}(\xi) \\ N_2(\xi)p_y^{1-2}(\xi) \\ \hline 0 \\ 0 \\ \hline 0 \\ 0 \end{array}\right]_{\xi,\eta=-1} t \underbrace{J}_{\frac{L^{1-2}}{2}} \mathrm{d}\xi + \int_{-1}^{+1} \left[\begin{array}{c} 0 \\ 0 \\ \hline N_2(\eta)p_x^{2-3}(\eta) \\ N_2(\eta)p_y^{2-3}(\eta) \\ \hline N_3(\eta)p_x^{2-3}(\eta) \\ N_3(\eta)p_y^{2-3}(v) \\ \hline 0 \\ 0 \end{array}\right]_{\xi=1,\eta} t \underbrace{J}_{\frac{L^{2-3}}{2}} \mathrm{d}\eta$$

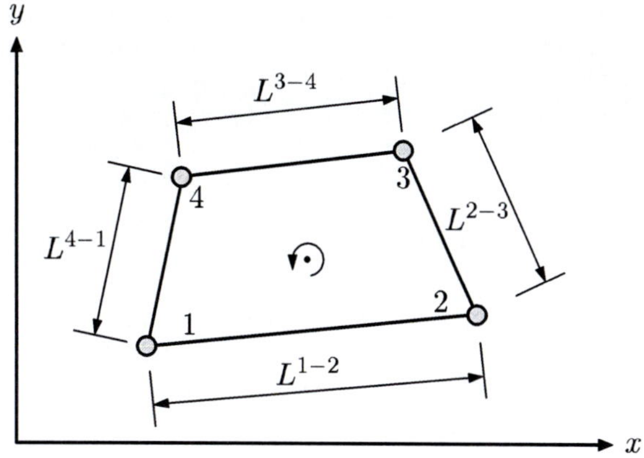

Fig. 3.17 Arbitrary four-node planar element in the Cartesian space (x, y) and definition of the edge lengths

$$
+\int_{+1}^{-1}\left[\begin{array}{c} 0 \\ 0 \\ \hline 0 \\ 0 \\ \hline N_3(\xi)p_x^{3-4}(\xi) \\ N_3(\xi)p_y^{3-4}(\xi) \\ \hline N_4(\xi)p_x^{3-4}(\xi) \\ N_4(\xi)p_y^{3-4}(\xi) \end{array}\right]_{\xi,\eta=1} t\underbrace{J}_{-\frac{L^{3-4}}{2}}\,\mathrm{d}\xi + \int_{+1}^{-1}\left[\begin{array}{c} N_1(\eta)p_x^{4-1}(\eta) \\ N_1(\eta)p_y^{4-1}(\eta) \\ \hline 0 \\ 0 \\ \hline 0 \\ 0 \\ \hline N_4(\eta)p_x^{4-1}(\eta) \\ N_4(\eta)p_y^{4-1}(\eta) \end{array}\right]_{\xi=-1,\eta} t\underbrace{J}_{-\frac{L^{4-1}}{2}}\,\mathrm{d}\eta\,. \tag{3.94}
$$

Let us recall here that the edge lengths can be calculated based on the nodal coordinates as:

$$L^{1-2} = \sqrt{(x_2 - x_1)^2 + (y_2 - y_1)^2}\,, \tag{3.95}$$

$$L^{2-3} = \sqrt{(x_3 - x_2)^2 + (y_3 - y_2)^2}\,, \tag{3.96}$$

$$L^{3-4} = \sqrt{(x_4 - x_3)^2 + (y_4 - y_3)^2}\,, \tag{3.97}$$

$$L^{4-1} = \sqrt{(x_1 - x_4)^2 + (y_1 - y_4)^2}\,. \tag{3.98}$$

Once the nodal displacements are known, e.g., based on $\boldsymbol{u}_{\mathrm{p}}^{\mathrm{e}} = (\boldsymbol{K}^{\mathrm{e}})^{-1}\boldsymbol{f}^{\mathrm{e}}$, further quantities can be calculated based on this result (the so-called post-processing, see Table 3.6).

Let us now have a closer look on the field equations provided in Table 3.6. The displacement field can be written as [5]

Table 3.6 Post-processing of nodal values for a four-node planar bilinear quadrilateral (quad 4). The distributions are given as being dependent on the nodal values $\boldsymbol{u}_{\mathrm{p}}^{\mathrm{e}}$ as a function of the natural coordinates $-1 \leq \xi \leq +1$ and $-1 \leq \eta \leq +1$

Displacement Field
$\boldsymbol{u}^{\mathrm{e}}(\xi,\eta) = \boldsymbol{N}^{\mathrm{T}}(\xi,\eta)\boldsymbol{u}_{\mathrm{p}}^{\mathrm{e}}$
Strain Field
$\boldsymbol{\varepsilon}^{\mathrm{e}}(\xi,\eta) = \boldsymbol{\mathcal{L}}_1\boldsymbol{u}^{\mathrm{e}}(\xi,\eta)$
Stress Field
$\boldsymbol{\sigma}^{\mathrm{e}}(\xi,\eta) = \boldsymbol{C}\boldsymbol{\mathcal{L}}_1\boldsymbol{u}^{\mathrm{e}}(\xi,\eta)$

$$\boldsymbol{u}^{\mathrm{e}} = \begin{bmatrix} u_x^{\mathrm{e}}(\xi,\eta) \\ u_y^{\mathrm{e}}(\xi,\eta) \end{bmatrix} = \left[\begin{array}{cc|c|c|cc} N_1(\xi,\eta) & 0 & \ldots & \ldots & N_4(\xi,\eta) & 0 \\ 0 & N_1(\xi,\eta) & \ldots & \ldots & 0 & N_4(\xi,\eta) \end{array}\right] \begin{bmatrix} u_{1x} \\ u_{1y} \\ u_{2x} \\ u_{2y} \\ u_{3x} \\ u_{3y} \\ u_{4x} \\ u_{4y} \end{bmatrix}, \tag{3.99}$$

where the interpolation functions $N_i(\xi,\eta)$ are given in Eqs. (3.12)–(3.15). The strain field can be written as [5]

$$\boldsymbol{\varepsilon}^{\mathrm{e}} = \begin{bmatrix} \varepsilon_x^{\mathrm{e}}(\xi,\eta) \\ \varepsilon_y^{\mathrm{e}}(\xi,\eta) \\ 2\varepsilon_{xy}^{\mathrm{e}}(\xi,\eta) \end{bmatrix} = \boldsymbol{\mathcal{L}}_1\boldsymbol{u}^{\mathrm{e}} = \begin{bmatrix} \frac{\partial}{\partial x} & 0 \\ 0 & \frac{\partial}{\partial y} \\ \frac{\partial}{\partial y} & \frac{\partial}{\partial x} \end{bmatrix} \begin{bmatrix} u_x^{\mathrm{e}}(\xi,\eta) \\ u_y^{\mathrm{e}}(\xi,\eta) \end{bmatrix} \tag{3.100}$$

$$= \left[\begin{array}{cc|c|c|cc} \frac{\partial N_1(\xi,\eta)}{\partial x} & 0 & \ldots & \ldots & \frac{\partial N_4(\xi,\eta)}{\partial x} & 0 \\ 0 & \frac{\partial N_1(\xi,\eta)}{\partial y} & \ldots & \ldots & 0 & \frac{\partial N_4(\xi,\eta)}{\partial y} \\ \frac{\partial N_1(\xi,\eta)}{\partial y} & \frac{\partial N_1(\xi,\eta)}{\partial x} & \ldots & \ldots & \frac{\partial N_4(\xi,\eta)}{\partial y} & \frac{\partial N_4(\xi,\eta)}{\partial x} \end{array}\right] \begin{bmatrix} u_{1x} \\ u_{1y} \\ u_{2x} \\ u_{2y} \\ u_{3x} \\ u_{3y} \\ u_{4x} \\ u_{4y} \end{bmatrix}, \tag{3.101}$$

where the partial derivatives must be calculated as follows (see Eqs. (3.16)–(3.19) and Eqs. (3.26)–(3.29) for intermediate results):

$$\frac{\partial N_i(\xi,\eta)}{\partial x} = \frac{\partial N_i(\xi,\eta)}{\partial \xi}\frac{\partial \xi}{\partial x} + \frac{\partial N_i(\xi,\eta)}{\partial \eta}\frac{\partial \eta}{\partial x}, \tag{3.102}$$

$$\frac{\partial N_i(\xi,\eta)}{\partial y} = \frac{\partial N_i(\xi,\eta)}{\partial \xi}\frac{\partial \xi}{\partial y} + \frac{\partial N_i(\xi,\eta)}{\partial \eta}\frac{\partial \eta}{\partial y}. \tag{3.103}$$

Once the strain field $\boldsymbol{\varepsilon}^{\mathrm{e}}(\xi, \eta)$ is obtained, multiplication with the elasticity matrix $\boldsymbol{C}$ gives the stress field $\boldsymbol{\sigma}^{\mathrm{e}}(\xi, \eta)$. It should be noted here that the strain and stress values are commonly evaluated in commercial finite element packages at the integration points. Different extrapolation strategies can be applied to project values to the nodes, which are used to display numerical values in a graphical interface (the so-called postprocessor).

3.2.2 Elements with Three Nodes

The principal finite element equation for a single linear three-node planar element can be stated as follows (see [5])

$$\begin{bmatrix} K_{11} & K_{12} & \dots & K_{15} & K_{16} \\ K_{21} & K_{22} & \dots & K_{25} & K_{26} \\ \vdots & \vdots & & \vdots & \vdots \\ K_{51} & K_{52} & \dots & K_{55} & K_{56} \\ K_{61} & K_{62} & \dots & K_{65} & K_{66} \end{bmatrix} \begin{bmatrix} u_{1x} \\ u_{1y} \\ u_{2x} \\ u_{2y} \\ u_{3x} \\ u_{3y} \end{bmatrix} = \int\limits_V \begin{bmatrix} N_1 & 0 \\ 0 & N_1 \\ \hline N_2 & 0 \\ 0 & N_2 \\ \hline N_3 & 0 \\ 0 & N_3 \end{bmatrix} \begin{bmatrix} f_x \\ f_y \end{bmatrix} \mathrm{d}V + \int\limits_S \begin{bmatrix} N_1 & 0 \\ 0 & N_1 \\ \hline N_2 & 0 \\ 0 & N_2 \\ \hline N_3 & 0 \\ 0 & N_3 \end{bmatrix} \begin{bmatrix} t_x \\ t_y \end{bmatrix} \mathrm{d}S\,, \tag{3.104}$$

or in abbreviated form

$$\boldsymbol{K}^{\mathrm{e}} \boldsymbol{u}_{\mathrm{p}}^{\mathrm{e}} = \boldsymbol{f}_{b}^{\mathrm{e}} + \boldsymbol{f}_{t}^{\mathrm{e}}\,, \tag{3.105}$$

where $\boldsymbol{K}^{\mathrm{e}}$ is the elemental stiffness matrix, $\boldsymbol{u}_{\mathrm{p}}^{\mathrm{e}}$ is the column matrix of nodal unknowns, $\boldsymbol{f}_{b}^{\mathrm{e}}$ is body force column matrix, and $\boldsymbol{f}_{t}^{\mathrm{e}}$ is the boundary force matrix.

Let us have now a closer look on the boundary force matrix $\boldsymbol{f}_{t}^{\mathrm{e}}$. The integral

$$\boldsymbol{f}_{t}^{\mathrm{e}} = \int\limits_S \boldsymbol{N} \boldsymbol{t} \, \mathrm{d}S\,, \tag{3.106}$$

which represents a surface integral, can be interpreted in different ways. One way is to understand it as single forces which are applied at nodes.[6] Then, we can write

[6] See [5] for further details.

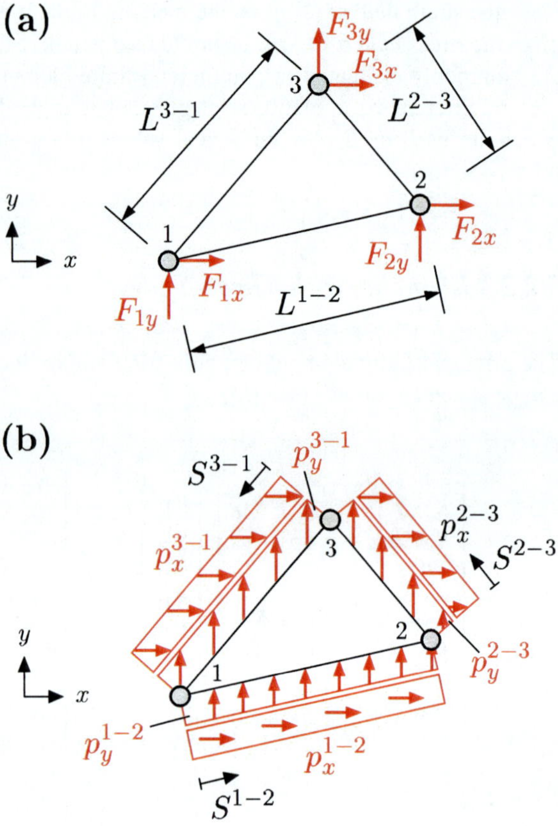

Fig. 3.18 Natural boundary conditions: **a** concentrated forces applied at single nodes, **b** distributed pressures at the edges of the element (each force and pressure is sketched in its positive direction)

Eq. (3.106) in the following way (see Fig. 3.18a):

$$\boldsymbol{f}_t^{\mathrm{e}} = \int_S \boldsymbol{N}\boldsymbol{t}\,\mathrm{d}S = \begin{bmatrix} F_{1x} \\ F_{1y} \\ F_{2x} \\ F_{2y} \\ F_{3x} \\ F_{3y} \end{bmatrix}. \tag{3.107}$$

Another view can be applied in the case of distributed pressures at the edges of the element, see Fig. 3.18b. Then we can write the surface integral in the following way:

$$\int_S \boldsymbol{N}\boldsymbol{t}\,\mathrm{d}S = \int_{1-2} \boldsymbol{N}|_{\xi,\eta,\zeta=0} \begin{bmatrix} p_x^{1-2}(x,y) \\ p_y^{1-2}(x,y) \end{bmatrix} \mathrm{d}S^{1-2} + \int_{2-3} \boldsymbol{N}|_{\xi=0,\eta,\zeta} \begin{bmatrix} p_x^{2-3}(x,y) \\ p_y^{2-3}(x,y) \end{bmatrix} \mathrm{d}S^{2-3}$$

$$
+ \int_{3-1} \boldsymbol{N}|_{\xi,\eta=0,\zeta} \begin{bmatrix} p_x^{3-1}(x,y) \\ p_y^{3-1}(x,y) \end{bmatrix} \mathrm{d}S^{3-1},
$$

$$
= \int_{1-2} \begin{bmatrix} N_1 & 0 \\ 0 & N_1 \\ \hline N_2 & 0 \\ 0 & N_2 \\ \hline 0 & 0 \\ 0 & 0 \end{bmatrix} \begin{bmatrix} p_x^{1-2}(x,y) \\ p_y^{1-2}(x,y) \end{bmatrix} \mathrm{d}S^{1-2} + \int_{2-3} \begin{bmatrix} 0 & 0 \\ 0 & 0 \\ \hline N_2 & 0 \\ 0 & N_2 \\ \hline N_3 & 0 \\ 0 & N_3 \end{bmatrix} \begin{bmatrix} p_x^{2-3}(x,y) \\ p_y^{2-3}(x,y) \end{bmatrix} \mathrm{d}S^{2-3}
$$

$$
+ \int_{3-1} \begin{bmatrix} 0 & 0 \\ 0 & 0 \\ \hline 0 & 0 \\ 0 & 0 \\ \hline N_3 & 0 \\ 0 & N_3 \end{bmatrix} \begin{bmatrix} p_x^{3-1}(x,y) \\ p_y^{3-1}(x,y) \end{bmatrix} \mathrm{d}S^{3-1}, \tag{3.108}
$$

or finally after the matrix multiplication:

$$
\int_S \boldsymbol{N} t \, \mathrm{d}S = \int_{1-2} \begin{bmatrix} N_1(\xi) p_x^{1-2}(x,y) \\ N_1(\xi) p_y^{1-2}(x,y) \\ \hline N_2(\xi) p_x^{1-2}(x,y) \\ N_2(\xi) p_y^{1-2}(x,y) \\ \hline 0 \\ 0 \end{bmatrix}_{\xi,\eta=1-\xi,\zeta=0} \mathrm{d}S^{1-2} + \int_{2-3} \begin{bmatrix} 0 \\ 0 \\ \hline N_2(\eta) p_x^{2-3}(x,y) \\ N_2(\eta) p_y^{2-3}(x,y) \\ \hline N_3(\eta) p_x^{2-3}(x,y) \\ N_3(\eta) p_y^{2-3}(x,y) \end{bmatrix}_{\xi=0,\eta,\zeta} \mathrm{d}S^{2-3}
$$

$$
+ \int_{3-1} \begin{bmatrix} N_1(\xi) p_x^{3-1}(x,y) \\ N_1(\xi) p_y^{3-1}(x,y) \\ \hline 0 \\ 0 \\ \hline N_3(\xi) p_x^{3-1}(x,y) \\ N_3(\xi) p_y^{3-1}(x,y) \end{bmatrix}_{\xi,\eta=0,\zeta} \mathrm{d}S^{3-1}. \tag{3.109}
$$

To transform the argument of the integration from $S = s \times t$ under the assumption of a constant thickness t to the natural coordinates ξ or η, we need to consider the relationships as indicated in Tables 3.7 and 3.8.

From Tables 3.7 and 3.8 we conclude the following relationships between natural and local coordinates:

Table 3.7 Relationship between natural coordinate ξ and local coordinate s^{1-2} (see Figs. 3.7 and 3.18b)

Node	Coordinate ξ	Coordinate s^{1-2}
1	+1	0
2	0	L^{1-2}

Table 3.8 Relationship between natural coordinate ξ and local coordinate s^{3-1} (see Fig. 3.7 and 3.18b)

Node	Coordinate ξ	Coordinate s^{3-1}
3	0	0
1	+1	L^{3-1}

$$s^{1-2} = (1-\xi)L^{1-2} \quad \text{or} \quad \mathrm{d}s^{1-2} = -L^{1-2}\mathrm{d}\xi\,, \tag{3.110}$$

$$s^{2-3} = (1-\eta)L^{2-3} \quad \text{or} \quad \mathrm{d}s^{2-3} = -L^{2-3}\mathrm{d}\eta\,, \tag{3.111}$$

$$s^{3-1} = \xi L^{3-1} \quad \text{or} \quad \mathrm{d}s^{3-1} = L^{3-1}\mathrm{d}\xi\,. \tag{3.112}$$

Thus, under consideration of the natural coordinates and the assumption that the thickness t is constant:

$$\int_S \boldsymbol{N} t\,\mathrm{d}S = \int_{+1}^{0} \begin{bmatrix} N_1(\xi)p_x^{1-2}(\xi) \\ N_1(\xi)p_y^{1-2}(\xi) \\ \hline N_2(\xi)p_x^{1-2}(\xi) \\ N_2(\xi)p_y^{1-2}(\xi) \\ \hline 0 \\ 0 \end{bmatrix} t \underbrace{J^{1-2}}_{-L^{1-2}} \mathrm{d}\xi + \int_{+1}^{0} \begin{bmatrix} 0 \\ 0 \\ \hline N_2(\eta)p_x^{2-3}(\eta) \\ N_2(\eta)p_y^{2-3}(\eta) \\ \hline N_3(\eta)p_x^{2-3}(\eta) \\ N_3(\eta)p_y^{2-3}(v) \end{bmatrix} t \underbrace{J^{2-3}}_{-L^{2-3}} \mathrm{d}\eta$$

$$+ \int_{0}^{+1} \begin{bmatrix} N_1(\xi)p_x^{3-1}(\xi) \\ N_1(\xi)p_y^{3-1}(\xi) \\ \hline 0 \\ 0 \\ \hline N_3(\xi)p_x^{3-1}(\xi) \\ N_3(\xi)p_y^{3-1}(\xi) \end{bmatrix} t \underbrace{J^{3-1}}_{L^{3-1}} \mathrm{d}\xi\,. \tag{3.113}$$

Let us recall here that the edge lengths can be calculated based on the nodal coordinates as:

$$L^{1-2} = \sqrt{(x_2-x_1)^2 + (y_2-y_1)^2}\,, \tag{3.114}$$

$$L^{2-3} = \sqrt{(x_3 - x_2)^2 + (y_3 - y_2)^2}\,, \tag{3.115}$$

$$L^{3-1} = \sqrt{(x_1 - x_3)^2 + (y_1 - y_3)^2}\,. \tag{3.116}$$

As we could see, the integration over triangular coordinates must be handled with care. In the case of a triangle with a constant thickness, the following integration rules can be applied for triangles with a constant Jacobian (e.g., straight sides and flat faces)[7]:

$$\int_s \xi^i \eta^j \mathrm{d}s = \frac{i!j!}{(i+j+a)!} \times L\,, \tag{3.117}$$

$$\int_0^1 \xi^i \eta^j \mathrm{d}\xi = \frac{i!j!}{(i+j+1)!}\,, \tag{3.118}$$

$$\int_0^1 \xi^i \eta^j \mathrm{d}\eta = \frac{i!j!}{(i+j+1)!}\,. \tag{3.119}$$

for $i \geq 0$ and $j \geq 0$.

3.2.3 Maxima Modules for the Finite Element Equation

3.2.4 Maxima Examples for the Principal Finite Element Equation

3.8 Example: Plate under tensile load

Given is a regular two-dimensional element as shown in Fig. 3.19. The left-hand nodes are fixed and the right-hand nodes are loaded by (a) horizontal forces F_0 or (b) a constant pressure $p_{x,0}$. Given are $a = 0.75$, $b = 0.5$ and $\nu = 0.2$ as numbers and F_0 or $p_{x,0}$, t, and E remain variable. Assume consistent units.

Use a single plane elasticity element to:

- Derive the general expression for the stiffness matrix under plane stress condition and for the right-hand side of the finite element equation.
- Calculate the nodal displacements as a functions of F_0 or $p_{x,0}$, t, and E. Use analytical and numerical 2×2 integration.

[7] Remember: $0! = 1$, $1! = 1$, $2! = 2$, and $3! = 6$.

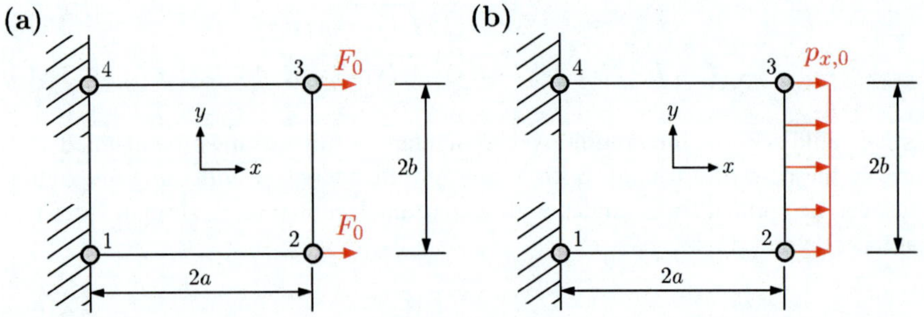

Fig. 3.19 Two-dimensional element under tensile load: **(a)** concentrated forces and **(b)** distributed pressure

3.8 Solution

The following Module 3.15 shows the entire wxMaxima code for the derivation of the stiffness matrix as well as the nodal displacements for the plane stress case with force boundary condition based on analytical integration.

The following Module 3.16 shows the entire wxMaxima code for the derivation of the stiffness matrix as well as the nodal displacements for the plane stress case with pressure boundary conditions based on analytical integration. It should be noted here that numerical 2×2 integration gives in both cases the same results.

```
→   kill(all);
    load("../Library/my_funs.mac")$
    fpprintprec : 6$ /* float numbers precision (only for printing) */
    ratprint : false$

(% o0)  done
```

Definition of Plane Stress Force Matrix: Nodal Coordinates "ncoor", Edges defining the Neumann boundaries "edges", Material Properties "mat", Number of Element Nodes "nen", Thickness Function "thickness", freedom tags list "nodtag", freedom values list "nodval", pressures on the edges "pressures"

```
→   a : 0.75$
    b : 0.5$

    ncoor : [[-a,-b],[a,-b],[a,b],[-a,b]]$

    thickness(x,y) := t$

    nodtag : [[1,1],[0,0],[0,0],[1,1]]$
    nodval : [[0,0],[F_0,0],[F_0,0],[0,0]]$
    edges : [[1,2],[2,3],[3,4],[4,1]]$
    pressures : [[0,0],[0,0],[0,0],[0,0]]$

    mat : [E, 0.2]$

    type : "plane stress"$
    nen : 4$
```

Maxima Calculations and Output

```
→   Ke : k_el_2D_plane_analytical(ncoor, mat, nen, type)$
    Fe : transpose(FlatNodePartVector(nodval))$
    Fe : Fe + f_edges_2D_plane_analytical(ncoor, pressures, nen, edges)$
    Ke_mod : ModifiedMasterStiffness(nodtag,Ke)$
    Fe_mod : ModifiedNodeForces(nodtag,nodval,Ke,Fe)$
    Ue : invert(Ke_mod).Fe_mod$

    print(" ")$
    print("=========")$
    print("Element stiffness matrix (analytical):")$
    print(" ", float(Ke))$

    print(" ")$
    print("=========")$
    print("Element force matrix (analytical):")$
    print(" ", float(Fe))$

    print(" ")$
    print("=========")$
    print("Modified element stiffness matrix (analytical):")$
    print(" ", float(Ke_mod))$

    print(" ")$
    print("=========")$
    print("Modified element force matrix (analytical):")$
    print(" ", float(Fe_mod))$

    print(" ")$
    print("=========")$
    print("Displacement matrix (analytical):")$
    print(" ", float(Ue))$
```

Element stiffness matrix (analytical):

$$\begin{bmatrix} 0.439815Et & 0.15625Et & -0.127315Et & -0.0520833Et & \ldots & 0.0520833Et \\ 0.15625Et & 0.613426Et & 0.0520833Et & 0.167824Et & \ldots & -0.474537Et \\ -0.127315Et & 0.0520833Et & 0.439815Et & -0.15625Et & \ldots & 0.15625Et \\ -0.0520833Et & 0.167824Et & -0.15625Et & 0.613426Et & \ldots & -0.306713Et \\ -0.219907Et & -0.15625Et & -0.0925926Et & 0.0520833Et & \ldots & -0.0520833Et \\ -0.15625Et & -0.306713Et & -0.0520833Et & -0.474537Et & \ldots & 0.167824Et \\ -0.0925926Et & -0.0520833Et & -0.219907Et & 0.15625Et & \ldots & -0.15625Et \\ 0.0520833Et & -0.474537Et & 0.15625Et & -0.306713Et & \ldots & 0.613426Et \end{bmatrix}$$

Element force matrix (analytical):

$$\begin{bmatrix} 0.0 \\ 0.0 \\ F_0 \\ 0.0 \\ F_0 \\ 0.0 \\ 0.0 \\ 0.0 \end{bmatrix}$$

Modified element stiffness matrix (analytical):

$$\begin{bmatrix} 1.0 & 0.0 & 0.0 & 0.0 & 0.0 & 0.0 & 0.0 & 0.0 \\ 0.0 & 1.0 & 0.0 & 0.0 & 0.0 & 0.0 & 0.0 & 0.0 \\ 0.0 & 0.0 & 0.439815Et & -0.15625Et & -0.0925926Et & -0.0520833Et & 0.0 & 0.0 \\ 0.0 & 0.0 & -0.15625Et & 0.613426Et & 0.0520833Et & -0.474537Et & 0.0 & 0.0 \\ 0.0 & 0.0 & -0.0925926Et & 0.0520833Et & 0.439815Et & 0.15625Et & 0.0 & 0.0 \\ 0.0 & 0.0 & -0.0520833Et & -0.474537Et & 0.15625Et & 0.613426Et & 0.0 & 0.0 \\ 0.0 & 0.0 & 0.0 & 0.0 & 0.0 & 0.0 & 1.0 & 0.0 \\ 0.0 & 0.0 & 0.0 & 0.0 & 0.0 & 0.0 & 0.0 & 1.0 \end{bmatrix}$$

Modified element force matrix (analytical):

$$\begin{bmatrix} 0.0 \\ 0.0 \\ F_0 \\ 0.0 \\ F_0 \\ 0.0 \\ 0.0 \\ 0.0 \end{bmatrix}$$

Displacement matrix (analytical):

$$\begin{bmatrix} 0.0 \\ 0.0 \\ \frac{2.96517F_0}{Et} \\ \frac{0.283899F_0}{Et} \\ \frac{2.96517F_0}{Et} \\ -\frac{0.283899F_0}{Et} \\ 0.0 \\ 0.0 \end{bmatrix}$$

Module 3.15: Stiffness matrices, element force matrices and displacement matrix for the plane stress case based with force boundary condition and analytical integration

```
→   kill(all);
    load("../Library/my_funs.mac")$
    fpprintprec : 6$ /* float numbers precision (only for printing) */
    ratprint : false$

(% o0)  done
```

Definition of Plane Stress Force Matrix: Nodal Coordinates "ncoor", Edges defining the Neumann boundaries "edges", Material Properties "mat", Number of Element Nodes "nen", Thickness Function "thickness", freedom tags list "nodtag", freedom values list "nodval", pressures on the edges "pressures"

```
→   a : 0.75$
    b : 0.5$

    ncoor : [[-a,-b],[a,-b],[a,b],[-a,b]]$

    thickness(x,y) := t$

    nodtag : [[1,1],[0,0],[0,0],[1,1]]$
    nodval : [[0,0],[0,0],[0,0],[0,0]]$
    edges : [[1,2],[2,3],[3,4],[4,1]]$
    pressures : [[0,0],[Px_0,0],[0,0],[0,0]]$

    mat : [E, 0.2]$

    type : "plane stress"$
    nen : 4$
```

Maxima Calculations and Output

```
→   Ke : k_el_2D_plane_analytical(ncoor, mat, nen, type)$
    Fe : transpose(FlatNodePartVector(nodval))$
    Fe : Fe + f_el_2D_plane_analytical(ncoor, pressures, nen, edges)$
    Ke_mod : ModifiedMasterStiffness(nodtag,Ke)$
    Fe_mod : ModifiedNodeForces(nodtag,nodval,Ke,Fe)$
    Ue : invert(Ke_mod).Fe_mod$

    print(" ")$
    print("=========")$
    print("Element stiffness matrix (analytical):")$
    print(" ", float(Ke))$

    print(" ")$
    print("=========")$
    print("Element force matrix (analytical):")$
    print(" ", float(Fe))$

    print(" ")$
    print("=========")$
    print("Modified element stiffness matrix (analytical):")$
    print(" ", float(Ke_mod))$

    print(" ")$
    print("=========")$
    print("Modified element force matrix (analytical):")$
    print(" ", float(Fe_mod))$

    print(" ")$
    print("=========")$
    print("Displacement matrix (analytical):")$
    print(" ", float(Ue))$
```

Element stiffness matrix (analytical):

$$\begin{bmatrix} 0.439815Et & 0.15625Et & -0.127315Et & -0.0520833Et & \ldots & 0.0520833Et \\ 0.15625Et & 0.613426Et & 0.0520833Et & 0.167824Et & \ldots & -0.474537Et \\ -0.127315Et & 0.0520833Et & 0.439815Et & -0.15625Et & \ldots & 0.15625Et \\ -0.0520833Et & 0.167824Et & -0.15625Et & 0.613426Et & \ldots & -0.306713Et \\ -0.219907Et & -0.15625Et & -0.0925926Et & 0.0520833Et & \ldots & -0.0520833Et \\ -0.15625Et & -0.306713Et & -0.0520833Et & -0.474537Et & \ldots & 0.167824Et \\ -0.0925926Et & -0.0520833Et & -0.219907Et & 0.15625Et & \ldots & -0.15625Et \\ 0.0520833Et & -0.474537Et & 0.15625Et & -0.306713Et & \ldots & 0.613426Et \end{bmatrix}$$

Element force matrix (analytical):

$$\begin{bmatrix} 0.0 \\ 0.0 \\ 0.5Px_0t \\ 0.0 \\ 0.5Px_0t \\ 0.0 \\ 0.0 \\ 0.0 \end{bmatrix}$$

Modified element stiffness matrix (analytical):

$$\begin{bmatrix} 1.0 & 0.0 & 0.0 & 0.0 & 0.0 & 0.0 & 0.0 & 0.0 \\ 0.0 & 1.0 & 0.0 & 0.0 & 0.0 & 0.0 & 0.0 & 0.0 \\ 0.0 & 0.0 & 0.439815Et & -0.15625Et & -0.0925926Et & -0.0520833Et & 0.0 & 0.0 \\ 0.0 & 0.0 & -0.15625Et & 0.613426Et & 0.0520833Et & -0.474537Et & 0.0 & 0.0 \\ 0.0 & 0.0 & -0.0925926Et & 0.0520833Et & 0.439815Et & 0.15625Et & 0.0 & 0.0 \\ 0.0 & 0.0 & -0.0520833Et & -0.474537Et & 0.15625Et & 0.613426Et & 0.0 & 0.0 \\ 0.0 & 0.0 & 0.0 & 0.0 & 0.0 & 0.0 & 1.0 & 0.0 \\ 0.0 & 0.0 & 0.0 & 0.0 & 0.0 & 0.0 & 0.0 & 1.0 \end{bmatrix}$$

Modified element force matrix (analytical):

$$\begin{bmatrix} 0.0 \\ 0.0 \\ 0.5Px_0t \\ 0.0 \\ 0.5Px_0t \\ 0.0 \\ 0.0 \\ 0.0 \end{bmatrix}$$

Displacement matrix (analytical):

$$\begin{bmatrix} 0.0 \\ 0.0 \\ \frac{1.48258Px_0}{E} \\ \frac{0.14195Px_0}{E} \\ \frac{1.48258Px_0}{E} \\ -\frac{0.14195Px_0}{E} \\ 0.0 \\ 0.0 \end{bmatrix}$$

Module 3.16: Stiffness matrices, element force matrices and displacement matrix for the plane stress case based with pressure boundary condition and analytical integration

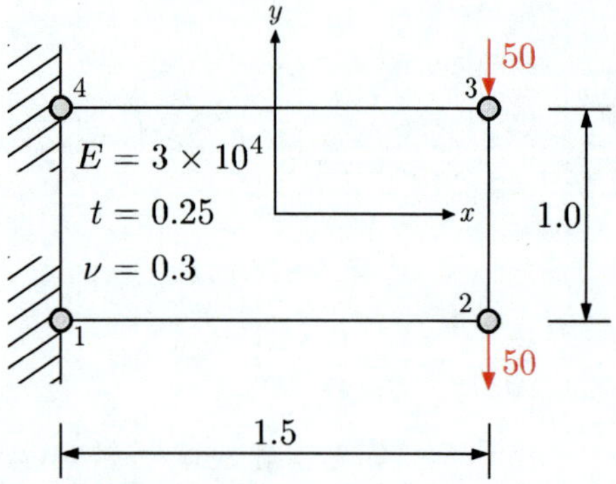

Fig. 3.20 Plane elasticity bending problem of a regular element

3.9 Example: Plane elasticity bending problem of a regular element

Given is a rectangular two-dimensional element (dimensions $1.5 \times 1.0 \times 0.25$) as shown in Fig. 3.20. The left-hand nodes are fixed and the right-hand nodes are loaded by vertical point loads $F_0 = -50$. Use a single plane elasticity element to calculate the nodal displacements for (a) a plane stress state and (b) a plane strain state with $E = 3 \times 10^4$ and $\nu = 0.3$. Assume consistent units. Base your computation on analytical integration.

3.9 Solution

The following Module 3.17 shows the entire wxMaxima code for the derivation of the stiffness matrix as well as the nodal displacements for the plane stress case with force boundary condition based on analytical integration.

The following Module 3.18 shows the entire wxMaxima code for the derivation of the stiffness matrix as well as the nodal displacements for the plane strain case with force boundary condition based on analytical integration.

```
-->   kill(all);
      load("../Library/my_funs.mac")$
      fpprintprec : 6$ /* float numbers precision (only for printing) */
      ratprint : false$

(% o0)  done
```

Definition of Plane Stress Force Matrix: Nodal Coordinates "ncoor", Edges defining the Neumann boundaries "edges", Material Properties "mat", Number of Element Nodes "nen", Thickness Function "thickness", freedom tags list "nodtag", freedom values list "nodval", pressures on the edges "pressures"

```
-->   a : 0.75$
      b : 0.5$

      ncoor : [[-a,-b],[a,-b],[a,b],[-a,b]]$

      thickness(x,y) := 0.25$
```

```
nodtag : [[1,1],[0,0],[0,0],[1,1]]$
nodval : [[0,0],[0,-50],[0,-50],[0,0]]$
edges : [[1,2],[2,3],[3,4],[4,1]]$
pressures : [[0,0],[0,0],[0,0],[0,0]]$

mat : [3.0E4, 0.3]$

type : "plane stress"$
nen : 4$
```

Maxima Calculations and Output

```
-->     Ke : k_el_2D_plane_analytical(ncoor, mat, nen, type)$
        Fe : transpose(FlatNodePartVector(nodval))$
        Fe : Fe + f_edges_2D_plane_analytical(ncoor, pressures, nen, edges)$
        Ke_mod : ModifiedMasterStiffness(nodtag,Ke)$
        Fe_mod : ModifiedNodeForces(nodtag,nodval,Ke,Fe)$
        Ue : invert(Ke_mod).Fe_mod$

        print(" ")$
        print("=========")$
        print("Element stiffness matrix (analytical):")$
        print(" ", float(Ke))$

        print(" ")$
        print("=========")$
        print("Element force matrix (analytical):")$
        print(" ", float(Fe))$

        print(" ")$
        print("=========")$
        print("Modified element stiffness matrix (analytical):")$
        print(" ", float(Ke_mod))$

        print(" ")$
        print("=========")$
        print("Modified element force matrix (analytical):")$
        print(" ", float(Fe_mod))$

        print(" ")$
        print("=========")$
        print("Displacement matrix (analytical):")$
        print(" ", float(Ue))$
```

Element stiffness matrix (analytical):

$$\begin{bmatrix} 3273.81 & 1339.29 & -1110.35 & -103.022 & -1636.9 & -1339.29 & -526.557 & 103.022 \\ 1339.29 & 4761.9 & 103.022 & 1419.41 & -1339.29 & -2380.95 & -103.022 & -3800.37 \\ -1110.35 & 103.022 & 3273.81 & -1339.29 & -526.557 & -103.022 & -1636.9 & 1339.29 \\ -103.022 & 1419.41 & -1339.29 & 4761.9 & 103.022 & -3800.37 & 1339.29 & -2380.95 \\ -1636.9 & -1339.29 & -526.557 & 103.022 & 3273.81 & 1339.29 & -1110.35 & -103.022 \\ -1339.29 & -2380.95 & -103.022 & -3800.37 & 1339.29 & 4761.9 & 103.022 & 1419.41 \\ -526.557 & -103.022 & -1636.9 & 1339.29 & -1110.35 & 103.022 & 3273.81 & -1339.29 \\ 103.022 & -3800.37 & 1339.29 & -2380.95 & -103.022 & 1419.41 & -1339.29 & 4761.9 \end{bmatrix}$$

Element force matrix (analytical):

$$\begin{bmatrix} 0.0 \\ 0.0 \\ 0.0 \\ -50.0 \\ 0.0 \\ -50.0 \\ 0.0 \\ 0.0 \end{bmatrix}$$

Modified element stiffness matrix (analytical):

$$\begin{bmatrix} 1.0 & 0.0 & 0.0 & 0.0 & 0.0 & 0.0 & 0.0 & 0.0 \\ 0.0 & 1.0 & 0.0 & 0.0 & 0.0 & 0.0 & 0.0 & 0.0 \\ 0.0 & 0.0 & 3273.81 & -1339.29 & -526.557 & -103.022 & 0.0 & 0.0 \\ 0.0 & 0.0 & -1339.29 & 4761.9 & 103.022 & -3800.37 & 0.0 & 0.0 \\ 0.0 & 0.0 & -526.557 & 103.022 & 3273.81 & 1339.29 & 0.0 & 0.0 \\ 0.0 & 0.0 & -103.022 & -3800.37 & 1339.29 & 4761.9 & 0.0 & 0.0 \\ 0.0 & 0.0 & 0.0 & 0.0 & 0.0 & 0.0 & 1.0 & 0.0 \\ 0.0 & 0.0 & 0.0 & 0.0 & 0.0 & 0.0 & 0.0 & 1.0 \end{bmatrix}$$

Modified element force matrix (analytical):

$$\begin{bmatrix} 0.0 \\ 0.0 \\ 0.0 \\ -50.0 \\ 0.0 \\ -50.0 \\ 0.0 \\ 0.0 \end{bmatrix}$$

Displacement matrix (analytical):

$$\begin{bmatrix} 0.0 \\ 0.0 \\ -0.0458182 \\ -0.120727 \\ 0.0458182 \\ -0.120727 \\ 0.0 \\ 0.0 \end{bmatrix}$$

Module 3.17: Stiffness matrices, element force matrices and displacement matrix for the plane stress case for a single element under bending load based on analytical integration

```
→  kill(all);
   load("../Library/my_funs.mac")$
   fpprintprec : 6$ /* float numbers precision (only for printing) */
   ratprint : false$

(% o0)  done
```

Definition of Plane Strain Force Matrix: Nodal Coordinates "ncoor", Edges defining the Neumann boundaries "edges", Material Properties "mat", Number of Element Nodes "nen", Thickness Function "thickness", freedom tags list "nodtag", freedom values list "nodval", pressures on the edges "pressures"

```
→  a : 0.75$
   b : 0.5$

   ncoor : [[-a,-b],[a,-b],[a,b],[-a,b]]$

   thickness(x,y) := 0.25$

   nodtag : [[1,1],[0,0],[0,0],[1,1]]$
   nodval : [[0,0],[0,-50],[0,-50],[0,0]]$
   edges : [[1,2],[2,3],[3,4],[4,1]]$
   pressures : [[0,0],[0,0],[0,0],[0,0]]$

   mat : [3.0E4, 0.3]$

   type : "plane strain"$
   nen : 4$
```

Maxima Calculations and Output

```
→  Ke : k_el_2D_plane_analytical(ncoor, mat, nen, type)$
   Fe : transpose(FlatNodePartVector(nodval))$
   Fe : Fe + f_edges_2D_plane_analytical(ncoor, pressures, nen, edges)$
   Ke_mod : ModifiedMasterStiffness(nodtag,Ke)$
   Fe_mod : ModifiedNodeForces(nodtag,nodval,Ke,Fe)$
   Ue : invert(Ke_mod).Fe_mod$

   print(" ")$
   print("=========")$
   print("Element stiffness matrix (analytical):")$
   print(" ", float(Ke))$

   print(" ")$
   print("=========")$
   print("Element force matrix (analytical):")$
   print(" ", float(Fe))$

   print(" ")$
   print("=========")$
   print("Modified element stiffness matrix (analytical):")$
   print(" ", float(Ke_mod))$

   print(" ")$
   print("=========")$
   print("Modified element force matrix (analytical):")$
   print(" ", float(Fe_mod))$

   print(" ")$
   print("=========")$
   print("Displacement matrix (analytical):")$
   print(" ", float(Ue))$
```

Element stiffness matrix (analytical):

$$\begin{bmatrix} 3685.9 & 1802.88 & -1522.44 & 360.577 & -1842.95 & -1802.88 & -320.513 & -360.577 \\ 1802.88 & 5689.1 & -360.577 & 1883.01 & -1802.88 & -2844.55 & 360.577 & -4727.56 \\ -1522.44 & -360.577 & 3685.9 & -1802.88 & -320.513 & 360.577 & -1842.95 & 1802.88 \\ 360.577 & 1883.01 & -1802.88 & 5689.1 & -360.577 & -4727.56 & 1802.88 & -2844.55 \\ -1842.95 & -1802.88 & -320.513 & -360.577 & 3685.9 & 1802.88 & -1522.44 & 360.577 \\ -1802.88 & -2844.55 & 360.577 & -4727.56 & 1802.88 & 5689.1 & -360.577 & 1883.01 \\ -320.513 & 360.577 & -1842.95 & 1802.88 & -1522.44 & -360.577 & 3685.9 & -1802.88 \\ -360.577 & -4727.56 & 1802.88 & -2844.55 & 360.577 & 1883.01 & -1802.88 & 5689.1 \end{bmatrix}$$

Element force matrix (analytical):

$$\begin{bmatrix} 0.0 \\ 0.0 \\ 0.0 \\ -50.0 \\ 0.0 \\ -50.0 \\ 0.0 \\ 0.0 \end{bmatrix}$$

Modified element stiffness matrix (analytical):

$$\begin{bmatrix} 1.0 & 0.0 & 0.0 & 0.0 & 0.0 & 0.0 & 0.0 & 0.0 \\ 0.0 & 1.0 & 0.0 & 0.0 & 0.0 & 0.0 & 0.0 & 0.0 \\ 0.0 & 0.0 & 3685.9 & -1802.88 & -320.513 & 360.577 & 0.0 & 0.0 \\ 0.0 & 0.0 & -1802.88 & 5689.1 & -360.577 & -4727.56 & 0.0 & 0.0 \\ 0.0 & 0.0 & -320.513 & -360.577 & 3685.9 & 1802.88 & 0.0 & 0.0 \\ 0.0 & 0.0 & 360.577 & -4727.56 & 1802.88 & 5689.1 & 0.0 & 0.0 \\ 0.0 & 0.0 & 0.0 & 0.0 & 0.0 & 0.0 & 1.0 & 0.0 \\ 0.0 & 0.0 & 0.0 & 0.0 & 0.0 & 0.0 & 0.0 & 1.0 \end{bmatrix}$$

Modified element force matrix (analytical):

$$\begin{bmatrix} 0.0 \\ 0.0 \\ 0.0 \\ -50.0 \\ 0.0 \\ -50.0 \\ 0.0 \\ 0.0 \end{bmatrix}$$

Displacement matrix (analytical):

$$\begin{bmatrix} 0.0 \\ 0.0 \\ -0.0406957 \\ -0.113043 \\ 0.0406957 \\ -0.113043 \\ 0.0 \\ 0.0 \end{bmatrix}$$

Module 3.18: Stiffness matrices, element force matrices and displacement matrix for the plane strain case for a single element under bending load based on analytical integration

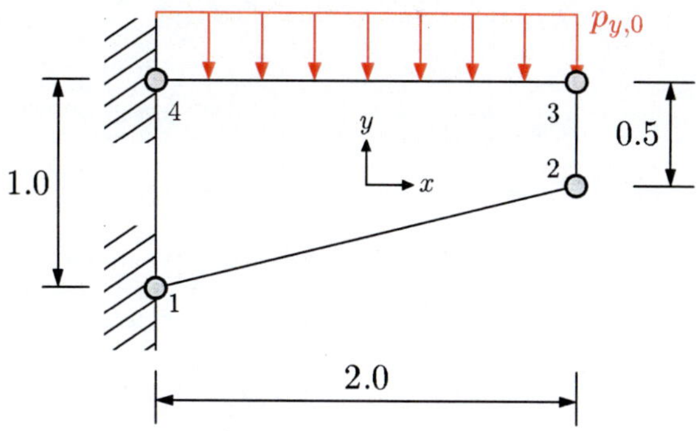

Fig. 3.21 Plane elasticity bending problem of a distorted element

3.10 Example: Plane elasticity bending problem of a distorted element

Given is a distorted two-dimensional element[8] as shown in Fig. 3.21. The constant thickness has a value of $t = 0.1$ and the material parameters are given as $E = 30000$ and $\nu = 0.3$. The upper horizontal edge is loaded by a constant pressure of $p_{y,0} = -20$, which is acting in the negative y-direction. Derive under the assumption of the plane stress condition the nodal displacements, as well as the strain and stress components on the integration points. Assume consistent units and analytical, numerical 2×2, 3×3, and 6×6 integration for the stiffness matrix $\boldsymbol{K}^{\mathrm{e}}$ and analytical integration for the boundary force matrix $\boldsymbol{f}_t^{\mathrm{e}}$.

3.10 Solution

The following Module 3.19 shows the entire wxMaxima code for the derivation of the stiffness matrix as well as the nodal displacements for the plane stress case of the distorted element based on analytical integration.

The following Module 3.20 shows the entire wxMaxima code for the derivation of the stiffness matrix as well as the nodal displacements for the plane stress case of the distorted element based on numerical 2×2 integration.

[8] This example is adapted from [2].

```
→     kill(all);
      load("../Library/my_funs.mac")$
      fpprintprec : 6$ /* float numbers precision (only for printing) */
      ratprint : false$

(% o0)  done
```

Definition of Plane Stress Force Matrix: Nodal Coordinates "ncoor", Edges defining the Neumann boundaries "edges", Material Properties "mat", Number of Element Nodes "nen", Thickness Function "thickness", freedom tags list "nodtag", freedom values list "nodval", pressures on the edges "pressures"

```
→     ncoor : [[-1.0,-0.5],[1.0,0.0],[1.0,0.5],[-1.0,0.5]]$

      thickness(x,y) := 0.1$

      nodtag : [[1,1],[0,0],[0,0],[1,1]]$
      nodval : [[0,0],[0,0],[0,0],[0,0]]$
      edges : [[3,4]]$
      pressures : [[0,-20]]$

      mat : [3.0E4, 0.3]$

      type : "plane stress"$
      nen : 4$
```

Maxima Calculations and Output

```
→     Ke : k_el_2D_plane_analytical(ncoor, mat, nen, type)$
      Fe : transpose(FlatNodePartVector(nodval))$
      Fe : Fe + f_edges_2D_plane_analytical(ncoor, pressures, nen, edges)$
      Ke_mod : ModifiedMasterStiffness(nodtag,Ke)$
      Fe_mod : ModifiedNodeForces(nodtag,nodval,Ke,Fe)$
      Ue : invert(Ke_mod).Fe_mod$

      print(" ")$
      print("=========")$
      print("Element stiffness matrix (analytical):")$
      print(" ", float(Ke))$

      print(" ")$
      print("=========")$
      print("Element force matrix (analytical):")$
      print(" ", float(Fe))$

      print(" ")$
      print("=========")$
      print("Modified element stiffness matrix (analytical):")$
      print(" ", float(Ke_mod))$

      print(" ")$
      print("=========")$
      print("Modified element force matrix (analytical):")$
      print(" ", float(Fe_mod))$

      print(" ")$
      print("=========")$
      print("Displacement matrix (analytical):")$
      print(" ", float(Ue))$
```

Element stiffness matrix (analytical):

$$\begin{bmatrix} 1081.87 & 328.771 & 143.945 & -163.036 & -556.033 & -413.887 & -669.786 & 248.152 \\ 328.771 & 2613.64 & -80.6187 & 1366.12 & -413.887 & -1510.35 & 165.735 & -2469.41 \\ 143.945 & -80.6187 & 2019.8 & -827.774 & -1195.63 & 250.851 & -968.121 & 657.542 \\ -163.036 & 1366.12 & -827.774 & 3861.17 & 333.268 & -3572.71 & 657.542 & -1654.58 \\ -556.033 & -413.887 & -1195.63 & 333.268 & 1607.71 & 243.655 & 143.945 & -163.036 \\ -413.887 & -1510.35 & 250.851 & -3572.71 & 243.655 & 3716.94 & -80.6187 & 1366.12 \\ -669.786 & 165.735 & -968.121 & 657.542 & 143.945 & -80.6187 & 1493.96 & -742.658 \\ 248.152 & -2469.41 & 657.542 & -1654.58 & -163.036 & 1366.12 & -742.658 & 2757.88 \end{bmatrix}$$

Element force matrix (analytical):

$$\begin{bmatrix} 0.0 \\ 0.0 \\ 0.0 \\ 0.0 \\ 0.0 \\ -2.0 \\ 0.0 \\ -2.0 \end{bmatrix}$$

Modified element stiffness matrix (analytical):

$$\begin{bmatrix} 1.0 & 0.0 & 0.0 & 0.0 & 0.0 & 0.0 & 0.0 & 0.0 \\ 0.0 & 1.0 & 0.0 & 0.0 & 0.0 & 0.0 & 0.0 & 0.0 \\ 0.0 & 0.0 & 2019.8 & -827.774 & -1195.63 & 250.851 & 0.0 & 0.0 \\ 0.0 & 0.0 & -827.774 & 3861.17 & 333.268 & -3572.71 & 0.0 & 0.0 \\ 0.0 & 0.0 & -1195.63 & 333.268 & 1607.71 & 243.655 & 0.0 & 0.0 \\ 0.0 & 0.0 & 250.851 & -3572.71 & 243.655 & 3716.94 & 0.0 & 0.0 \\ 0.0 & 0.0 & 0.0 & 0.0 & 0.0 & 0.0 & 1.0 & 0.0 \\ 0.0 & 0.0 & 0.0 & 0.0 & 0.0 & 0.0 & 0.0 & 1.0 \end{bmatrix}$$

Modified element force matrix (analytical):

$$\begin{bmatrix} 0.0 \\ 0.0 \\ 0.0 \\ 0.0 \\ 0.0 \\ -2.0 \\ 0.0 \\ 0.0 \end{bmatrix}$$

Displacement matrix (analytical):

$$\begin{bmatrix} 0.0 \\ 0.0 \\ -0.00114926 \\ -0.00955005 \\ 0.0026119 \\ -0.00981121 \\ 0.0 \\ 0.0 \end{bmatrix}$$

Module 3.19: Stiffness matrices, element force matrices and displacement matrix for the plane stress case for the distorted single element based on analytical integration

```
→   kill(all);
    load("../Library/my_funs.mac")$
    fpprintprec : 6$ /* float numbers precision (only for printing) */
    ratprint :false$

(% o0)  done
```

Definition of Plane Stress Force Matrix: Nodal Coordinates "ncoor", Edges defining the Neumann boundaries "edges", Material Properties "mat", Number of Element Nodes "nen", Thickness Function "thickness", freedom tags list "nodtag", freedom values list "nodval", pressures on the edges "pressures"

```
→   ncoor : [[-1.0,-0.5],[1.0,0.0],[1.0,0.5],[-1.0,0.5]]$

    thickness(x,y) := 0.1$

    nodtag : [[1,1],[0,0],[0,0],[1,1]]$
    nodval : [[0,0],[0,0],[0,0],[0,0]]$
    edges : [[3,4]]$
    pressures : [[0,-20]]$

    mat : [3.0E4, 0.3]$

    type : "plane stress"$
    nen : 4$
    nqp : 4$
```

Maxima Calculations and Output

```
→   Ke : k_el_2D_plane_numerical(ncoor, mat, nqp, nen, type)$
    Fe : transpose(FlatNodePartVector(nodval))$
    Fe : Fe + f_edges_2D_plane_analytical(ncoor, pressures, nen, edges)$
    Ke_mod : ModifiedMasterStiffness(nodtag,Ke)$
    Fe_mod : ModifiedNodeForces(nodtag,nodval,Ke,Fe)$
    Ue : invert(Ke_mod).Fe_mod$
    strains : strain_el_2D_plane_numerical(ncoor, nqp, nen, Ue, type)$
    stresses: stress_el_2D_plane_numerical(mat, nqp, strains, type)$

    print(" ")$
    print("=========")$
    print("Element stiffness matrix (numerical):")$
    print(" ", float(Ke))$

    print(" ")$
    print("=========")$
    print("Element force matrix (analytical):")$
    print(" ", float(Fe))$

    print(" ")$
    print("=========")$
    print("Modified element stiffness matrix (numerical):")$
    print(" ", float(Ke_mod))$

    print(" ")$
    print("=========")$
```

```
print("Modified element force matrix (analytical):")$
print(" ", float(Fe_mod))$

print(" ")$
print("=========")$
print("Displacement matrix:")$
print(" ", float(Ue))$

print(" ")$
print("=========")$
print("Strains and stresses on quadrature points:")$
strain_symb : transpose([_xx,_yy,2*_xy])$
stress_symb : transpose([_xx,_yy,_xy])$
print(" ")$
for l : 1 thru nqp do(
        print(" ", "Quadrature point:", l),
        print(" ", strain_symb, "=", float(strains[l]), " ", stress_symb, "=", float(stresses[l])),
        print(" ")
)$
```

Element stiffness matrix (numerical):

$$\begin{bmatrix} 1077.77 & 329.67 & 152.156 & -164.835 & -564.243 & -412.088 & -665.68 & 247.253 \\ 329.67 & 2602.49 & -82.4176 & 1388.42 & -412.088 & -1532.65 & 164.835 & -2458.26 \\ 152.156 & -82.4176 & 2003.38 & -824.176 & -1179.21 & 247.253 & -976.331 & 659.341 \\ -164.835 & 1388.42 & -824.176 & 3816.57 & 329.67 & -3528.11 & 659.341 & -1676.88 \\ -564.243 & -412.088 & -1179.21 & 329.67 & 1591.29 & 247.253 & 152.156 & -164.835 \\ -412.088 & -1532.65 & 247.253 & -3528.11 & 247.253 & 3672.34 & -82.4176 & 1388.42 \\ -665.68 & 164.835 & -976.331 & 659.341 & 152.156 & -82.4176 & 1489.86 & -741.758 \\ 247.253 & -2458.26 & 659.341 & -1676.88 & -164.835 & 1388.42 & -741.758 & 2746.72 \end{bmatrix}$$

Element force matrix (analytical):

$$\begin{bmatrix} 0.0 \\ 0.0 \\ 0.0 \\ 0.0 \\ 0.0 \\ -2.0 \\ 0.0 \\ -2.0 \end{bmatrix}$$

Modified element stiffness matrix (numerical):

$$\begin{bmatrix} 1.0 & 0.0 & 0.0 & 0.0 & 0.0 & 0.0 & 0.0 & 0.0 \\ 0.0 & 1.0 & 0.0 & 0.0 & 0.0 & 0.0 & 0.0 & 0.0 \\ 0.0 & 0.0 & 2003.38 & -824.176 & -1179.21 & 247.253 & 0.0 & 0.0 \\ 0.0 & 0.0 & -824.176 & 3816.57 & 329.67 & -3528.11 & 0.0 & 0.0 \\ 0.0 & 0.0 & -1179.21 & 329.67 & 1591.29 & 247.253 & 0.0 & 0.0 \\ 0.0 & 0.0 & 247.253 & -3528.11 & 247.253 & 3672.34 & 0.0 & 0.0 \\ 0.0 & 0.0 & 0.0 & 0.0 & 0.0 & 0.0 & 1.0 & 0.0 \\ 0.0 & 0.0 & 0.0 & 0.0 & 0.0 & 0.0 & 0.0 & 1.0 \end{bmatrix}$$

Modified element force matrix (analytical):

$$\begin{bmatrix} 0.0 \\ 0.0 \\ 0.0 \\ 0.0 \\ 0.0 \\ -2.0 \\ 0.0 \\ 0.0 \end{bmatrix}$$

Displacement matrix:

$$\begin{bmatrix} 0.0 \\ 0.0 \\ -0.00117777 \\ -0.00966972 \\ 0.00267425 \\ -0.00993532 \\ 0.0 \\ 0.0 \end{bmatrix}$$

==========

Strains and stresses on quadrature points:

Quadrature point: 1

$$\begin{bmatrix} \epsilon_{xx} \\ \epsilon_{yy} \\ 2\epsilon_{xy} \end{bmatrix} = \begin{bmatrix} -3.61335\times10^{-4} \\ -6.27569\times10^{-5} \\ -0.00394035 \end{bmatrix} \quad \begin{bmatrix} \sigma_{xx} \\ \sigma_{yy} \\ \sigma_{xy} \end{bmatrix} = \begin{bmatrix} -12.5328 \\ -5.64255 \\ -45.4656 \end{bmatrix}$$

Quadrature point: 2

$$\begin{bmatrix} \epsilon_{xx} \\ \epsilon_{yy} \\ 2\epsilon_{xy} \end{bmatrix} = \begin{bmatrix} -0.00117087 \\ -3.45844\times10^{-4} \\ 2.21253\times10^{-4} \end{bmatrix} \quad \begin{bmatrix} \sigma_{xx} \\ \sigma_{yy} \\ \sigma_{xy} \end{bmatrix} = \begin{bmatrix} -42.0205 \\ -22.9815 \\ 2.55292 \end{bmatrix}$$

Quadrature point: 3

$$\begin{bmatrix} \epsilon_{xx} \\ \epsilon_{yy} \\ 2\epsilon_{xy} \end{bmatrix} = \begin{bmatrix} 6.65111\times10^{-4} \\ -3.45844\times10^{-4} \\ 9.46655\times10^{-5} \end{bmatrix} \quad \begin{bmatrix} \sigma_{xx} \\ \sigma_{yy} \\ \sigma_{xy} \end{bmatrix} = \begin{bmatrix} 18.5063 \\ -4.82341 \\ 1.09229 \end{bmatrix}$$

Quadrature point: 4

$$\begin{bmatrix} \epsilon_{xx} \\ \epsilon_{yy} \\ 2\epsilon_{xy} \end{bmatrix} = \begin{bmatrix} 8.82025\times10^{-4} \\ -6.27569\times10^{-5} \\ -0.00402608 \end{bmatrix} \quad \begin{bmatrix} \sigma_{xx} \\ \sigma_{yy} \\ \sigma_{xy} \end{bmatrix} = \begin{bmatrix} 28.4571 \\ 6.65441 \\ -46.4547 \end{bmatrix}$$

Module 3.20: Stiffness matrices, element force matrices and displacement matrix for the plane stress case for the distorted single element based on numerical 2×2 integration

The difference in the results for the displacement matrix depending on the integration procedures is shown in Table 3.9.

3.11 Advanced Example: Different plane modeling approaches of a simply supported beam

Given is a simply supported Euler-Bernoulli beam as indicated Fig. 3.22. The length of the beam is $4a$ and the rectangular cross section has the dimensions $t \times 2b$. The beam is loaded by a single force F_0 acting in the middle of the beam. Note that the problem is *not* symmetric.

Use plane elasticity two-dimensional elements in the following to model the problem and to calculate the nodal unknowns at node 2 under the assumption of a plane *stress* case. The modelling approach is based on two elements with nodes 1, ..., 6, see Fig. 3.23.

Table 3.9 Comparison of the results for the displacement matrix depending on the integration procedure

Analytical	2 × 2	3 × 3	6 × 6
0.0	0.0	0.0	0.0
0.0	0.0	0.0	0.0
−0.00114926	−0.00117777	−0.0011501	−0.00114926
−0.00955005	−0.00966972	−0.0095536	−0.00955005
0.0026119	0.00267425	0.00261375	0.0026119
−0.00981121	−0.00993532	−0.00981489	−0.00981121
0.0	0.0	0.0	0.0
0.0	0.0	0.0	0.0

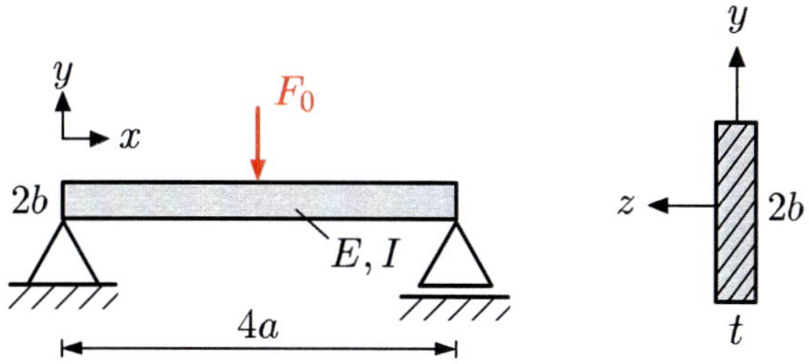

Fig. 3.22 Simply supported beam with rectangular cross section

Case (a) Calculate the symbolic solution for the nodal unknowns (displacements) at node 2 under the assumption that the force F_0 is acting at node 3.

Case (b) Calculate the symbolic solution for nodal unknowns (displacements) at node 2 under the assumption that the force F_0 is acting at node 2.

Case (c) Calculate the symbolic solution for the nodal unknowns (displacements) at node 2 under the assumption that the force $\frac{1}{2}F_0$ is acting at node 3 and $\frac{1}{2}F_0$ is acting at node 2.

Compare the results for the vertical displacement at node 2 for the special case of $a = 0.75$, $b = 0.5$ and $\nu = 0.2$.

Case (d) Calculate the numerical solution ($a = 0.75, b = 0.5, \nu = 0.2$ and $t = 0.2a$) for the nodal unknowns (displacements) at node 2 under the assumption that the force F_0 is acting at node 2. Pay attention to the fact that the elements are no longer rectangular.

3.11 Solution

The following Module 3.21 shows the entire wxMaxima code for the load case (a), i.e. the force F_0 is acting at node 3.

The following Module 3.22 shows the entire wxMaxima code for the load case (d), i.e. the force F_0 is acting at node 2 and distorted elements.

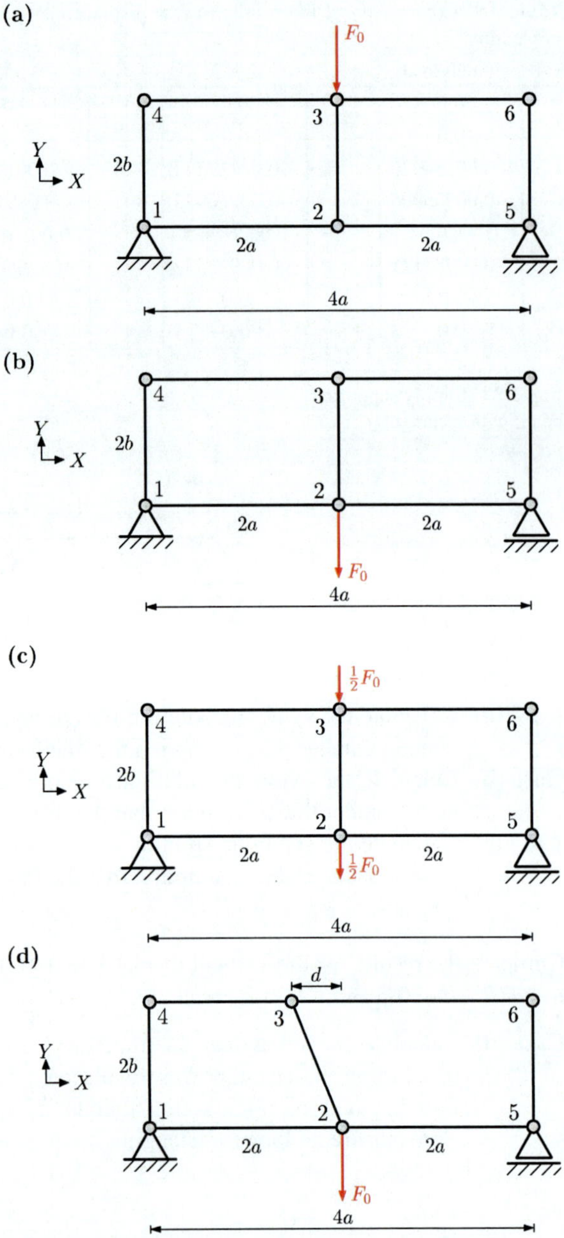

Fig. 3.23 Different modeling approaches for the beam shown in Fig. 3.22

```
→   kill(all);
    load("../Library/my_funs.mac")$
    fpprintprec : 6$ /* float numbers precision (only for printing) */
    ratprint :false$

(% o0) done
```

Definition of Plane Stress Elasticity Problem: Nodal Coordinates "nodxy", Element Connectivity "elenod", Element Material Properties "elemat", Number of Quadrature Points "nqp", Number of Element Nodes "nen", Thickness Function "thickness", freedom tags list "nodtag", freedom values list "nodval", Edges defining the Neumann boundaries "edges", pressures on the edges "pressures", Numerical or Analytical Calculations Switch "numerical"

```
→   a : 0.75$
    b : 0.5$

    nodxy : [[0,-b],[2*a,-b],[2*a,b],[0,b],[4*a,-b],[4*a,b]]$
    elenod : [[1,2,3,4],[2,5,6,3]]$
    elemat : [[E, 0.2],[E, 0.2]]$

    numnod : length(nodxy)$
    numele : length(elenod)$

    thickness(x,y) := t$

    nodtag : makelist([0,0],numnod)$
    nodval : makelist([0,0],numnod)$
    nodtag[1] : [1,1]$
    nodtag[5] : [0,1]$
    nodval[3] : [0,-F_0]$
    edges : [[1,2,5],[5,6],[6,3,4],[4,1]]$
    pressures : [[0,0],[0,0],[0,0],[0,0]]$
    body : [0,0]$

    type : "plane stress"$
    numerical : true $
    nen : 4$
    nqp : 4$
```

Maxima Calculations and Output

```
→   [noddis,~,~] : PlanarElementsSolution(nodxy,elenod,elemat,edges,pressures,body,
                   nodtag,nodval,numerical)$

    print(" ")$
    print("Displacements at Node 2:")$
    print(" ", u_2X, "=", float(noddis[2][1]))$
    print(" ", u_2Y, "=", float(noddis[2][2]))$
```

Displacements at Node 2:

$$u_{2X} = \frac{1.80526 F_0}{E t}$$

$$u_{2Y} = -\frac{4.35789 F_0}{E t}$$

Module 3.21: Simply supported beam solution for the load case (a), i.e., the force F_0 is acting at node 3

```
→    kill(all);
     load("../Library/my_funs.mac")$
     fpprintprec : 6$ /* float numbers precision (only for printing) */
     ratprint :false$

(% o0)   done
```

Definition of Plane Stress Elasticity Problem: Nodal Coordinates "nodxy", Element Connectivity "elenod", Element Material Properties "elemat", Number of Quadrature Points "nqp", Number of Element Nodes "nen", Thickness Function "thickness", freedom tags list "nodtag", freedom values list "nodval", Edges defining the Neumann boundaries "edges", pressures on the edges "pressures", Numerical or Analytical Calculations Switch "numerical"

```
→    a : 0.75$
     b : 0.5$
     d : 0.2*a$

     nodxy : [[0,-b],[2*a,-b],[2*a-d,b],[0,b],[4*a,-b],[4*a,b]]$
     elenod : [[1,2,3,4],[2,5,6,3]]$
     elemat : [[E, 0.2],[E, 0.2]]$

     numnod : length(nodxy)$
     numele : length(elenod)$

     thickness(x,y) := t$

     nodtag : makelist([0,0],numnod)$
     nodval : makelist([0,0],numnod)$
     nodtag[1] : [1,1]$
     nodtag[5] : [0,1]$
```

```
nodval[2] : [0,-F_0]$
edges : [[1,2,5],[5,6],[6,3,4],[4,1]]$
pressures : [[0,0],[0,0],[0,0],[0,0]]$
body : [0,0]$

type : "plane stress"$
numerical : true $
nen : 4$
nqp : 9$
```

Maxima Calculations and Output

```
→   [noddis,~,~] : PlanarElementsSolution(nodxy,elenod,elemat,edges,pressures,body,
                   nodtag,nodval,numerical)$

    print(" ")$
    print("Displacements at Node 2:")$
    print(" ", u_2X, "=", float(noddis[2][1]))$
    print(" ", u_2Y, "=", float(noddis[2][2]))$
```

Displacements at Node 2:

$$u_{2X} = \frac{1.79524 F_0}{Et}$$

$$u_{2Y} = -\frac{4.99679 F_0}{Et}$$

Module 3.22: Simply supported beam solution for the load case (d), i.e., the force F_0 is acting at node 2 and distorted elements

The different results for the displacements at note 2 are shown in Table 3.10.

3.12 Advanced Example: Simply supported beam with distributed load

Given is a finite element representation of simply supported beam as indicated Fig. 3.24. The length of the beam is $4a$ and the rectangular cross section has the dimensions $t \times 2b$. The beam is loaded by a constant pressure p_0 acting in the upper edge of the beam.

Table 3.10 Comparison of the results for the displacements at node 2 depending on the modeling approach

Case	Integration	$u_{2X} / \frac{F_0}{Et}$	$u_{2Y} / \frac{F_0}{Et}$
(a)	2×2	1.80526	−4.35789
(b)	2×2	1.70526	−5.17299
(c)	2×2	1.75526	−4.76544
(d)	2×2	1.79589	−4.99772
	3×3	1.79524	−4.99679
	analyt.	1.79524	−4.99679

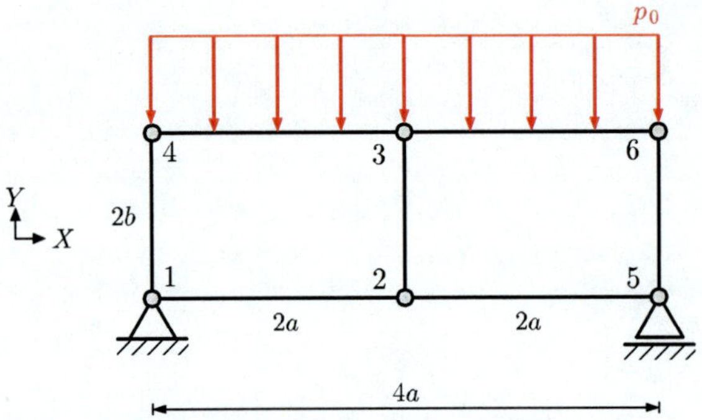

Fig. 3.24 Simply supported beam with constant pressure

Use two plane elasticity elements to calculate the nodal unknowns at node 2 under the assumption of a plane *stress* case and a magnitude of the pressure of $|p_0| = \frac{F_0}{2at}$. Base your solution on analytical integration.

3.12 Solution

The following Module 3.23 shows the entire wxMaxima code for the simply supported beam with constant pressure.

```
kill(all);
load("../Library/my_funs.mac")$
fpprintprec : 6$ /* float numbers precision (only for printing) */
ratprint :false$

(% o0)  done
```

Definition of Plane Stress Elasticity Problem: Nodal Coordinates "nodxy", Element Connectivity "elenod", Element Material Properties "elemat", Number of Quadrature Points "nqp", Number of Element Nodes "nen", Thickness Function "thickness", freedom tags list "nodtag", freedom values list "nodval", Edges defining the Neumann boundaries "edges", pressures on the edges "pressures", Numerical or Analytical Calculations Switch "numerical"

```
a : 0.75$
b : 0.5$
p_0 : F_0/(2*a*t)$

nodxy : [[0,-b],[2*a,-b],[2*a,b],[0,b],[4*a,-b],[4*a,b]]$
elenod : [[1,2,3,4],[2,5,6,3]]$
elemat : [[E, 0.2],[E, 0.2]]$

numnod : length(nodxy)$
numele : length(elenod)$
```

```
nodtag : makelist([0,0],numnod)$
nodval : makelist([0,0],numnod)$
nodtag[1] : [1,1]$
nodtag[5] : [0,1]$
edges : [[1,2,5],[5,6],[6,3,4],[4,1]]$
pressures : [[0,0],[0,0],[0,-p_0],[0,0]]$
body : [0,0]$

type : "plane stress"$
numerical : false $
nen : 4$
```

Maxima Calculations and Output

```
→   [noddis,~,~] : PlanarElementsSolution(nodxy,elenod,elemat,edges,pressures,body,
                     nodtag,nodval,numerical)$

    print(" ")$
    print("Displacements at Node 2:")$
    print(" ", u_2X, "=", float(noddis[2][1]))$
    print(" ", u_2Y, "=", float(noddis[2][2]))$
```

Displacements at Node 2:

$$u_{2X} = \frac{1.90526 F_0}{Et}$$

$$u_{2Y} = -\frac{5.17299 F_0}{Et}$$

Module 3.23: Simply supported beam solution with constant pressure

3.13 Example: Plane elasticity problem of a three-node element with distributed load: boundary force matrix

Given is a three-node plane elasticity element as shown in Fig. 3.25. The constant thickness has a value of t and the materials parameters are given as E and ν. The lower edge $1-2$ is loaded by a linearly changing pressure (value at node 1: $p_{x,1}^{1-2}$; value at node 2: $p_{x,2}^{1-2}$), which is acting in the positive x-direction. Derive based on analytical integration the expression for the boundary force matrix $\boldsymbol{f}_t^{\mathrm{e}}$ as a function of the edge length L^{1-2}. Simplify your general result for the special cases (a) $p_{x,1}^{1-2} = p_{x,2}^{1-2} = p_x^{1-2} = \text{const.}$ and (b) $p_{x,1}^{1-2} = 0$.

3.13 Solution

The evaluation of Eq. (3.113) gives under consideration of

$$p_x^{1-2}(\xi) = p_{x,1}^{1-2}\xi + p_{x,2}^{1-2}(1-\xi)\,, \tag{3.120}$$

$$N_1(\xi) = \xi\,, \tag{3.121}$$

$$N_2(\xi) = 1-\xi\,, \tag{3.122}$$

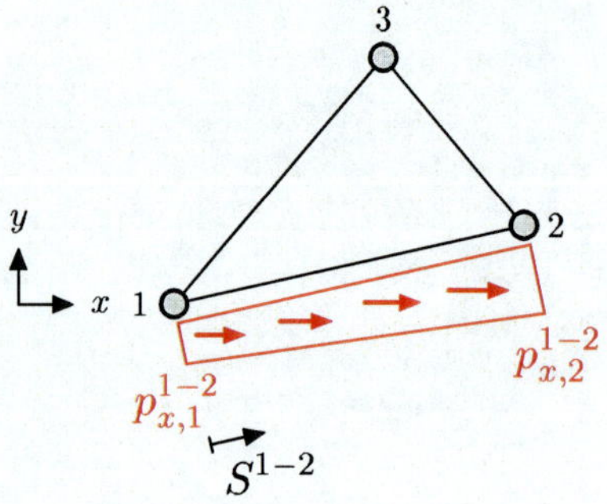

Fig. 3.25 Plane elasticity problem of a three-node element: boundary force

the following expression:

$$\int_S \boldsymbol{N} t \, \mathrm{d}S = \int_{\xi=+1}^{\xi=0} \begin{bmatrix} N_1(\xi) p_x^{1-2}(\xi) \\ 0 \\ \hline N_2(\xi) p_x^{1-2}(\xi) \\ 0 \\ \hline 0 \\ 0 \end{bmatrix} t \underbrace{J^{1-2}}_{-L^{1-2}} \mathrm{d}\xi$$

$$= \int_{+1}^{0} \begin{bmatrix} \xi \left\{ p_{x,1}^{1-2}\xi + p_{x,2}^{1-2}(1-\xi) \right\} \\ 0 \\ \hline (1-\xi) \left\{ p_{x,1}^{1-2}\xi + p_{x,2}^{1-2}(1-\xi) \right\} \\ 0 \\ \hline 0 \\ 0 \end{bmatrix} t(-L^{1-2}) \mathrm{d}\xi \tag{3.123}$$

$$= tL^{1-2} \begin{bmatrix} \frac{1}{3} p_{x,1}^{1-2} + \frac{1}{6} p_{x,2}^{1-2} \\ 0 \\ \hline \frac{1}{6} p_{x,1}^{1-2} + \frac{1}{3} p_{x,2}^{1-2} \\ 0 \\ \hline 0 \\ 0 \end{bmatrix} . \tag{3.124}$$

The special case $p_{x,1}^{1-2} = p_{x,2}^{1-2} = p_x^{1-2} = \text{const.}$ gives

$$\int_S \boldsymbol{N} t \, \mathrm{d}S = t L^{1-2} p_x^{1-2} \begin{bmatrix} \frac{1}{2} \\ 0 \\ \hline \frac{1}{2} \\ 0 \\ \hline 0 \\ 0 \end{bmatrix}, \tag{3.125}$$

whereas the special case $p_{x,1}^{1-2} = 0$ gives:

$$\int_S \boldsymbol{N} t \, \mathrm{d}S = t L^{1-2} p_{x,2}^{1-2} \begin{bmatrix} \frac{1}{6} \\ 0 \\ \hline \frac{1}{3} \\ 0 \\ \hline 0 \\ 0 \end{bmatrix}. \tag{3.126}$$

The following Module 3.24 shows the entire wxMaxima code for the boundary force matrix.

```
-->   kill(all);
      load("../Library/my_funs.mac")$
      fpprintprec : 6$ /* float numbers precision (only for printing) */
      ratprint :false$

(%o0) done
```

Definition of Plane Strain Force Matrix: Nodal Coordinates "ncoor", Edges defining the Neumann boundaries "edges", Number of Element Nodes "nen", Thickness Function "thickness", pressures on the edges "pressures"

```
-->   ncoor : [[x1,y1],[x2,y2],[x3,y3]]$

      thickness(x,y) := t$

      edges : [[1,2]]$

      /* general case */
      pressures[1] : [[[p_x1,0],[p_x2,0]]]$

      /* case (a) */
      pressures[2] : [[[p_x,0],[p_x,0]]]$
      /* case (b) */
      pressures[3] : [[[0,0],[p_x_2,0]]]$

      nen : 3$
```

Maxima Calculations and Output

```
→    print(" ")$
     print("General case :")$
     f_el : triangular_f_edges_2D_plane_analytical(ncoor, pressures[1], nen, edges)$
     print(" f = ", f_el)$
     print(" ")$
     print("Case (a) :")$
     f_el : triangular_f_edges_2D_plane_analytical(ncoor, pressures[2], nen, edges)$
     print(" f = ", f_el)$
     print(" ")$
     print("Case (b) :")$
     f_el : triangular_f_edges_2D_plane_analytical(ncoor, pressures[3], nen, edges)$
     print(" f = ", f_el)$
```

General case :

$$f = \begin{pmatrix} \frac{(p_{x2}+2p_{x1})t\sqrt{(y2-y1)^2+(x2-x1)^2}}{6} \\ 0 \\ \frac{(2p_{x2}+p_{x1})t\sqrt{(y2-y1)^2+(x2-x1)^2}}{6} \\ 0 \\ 0 \\ 0 \end{pmatrix}$$

Case (a) :

$$f = \begin{pmatrix} \frac{p_x t\sqrt{(y2-y1)^2+(x2-x1)^2}}{2} \\ 0 \\ \frac{p_x t\sqrt{(y2-y1)^2+(x2-x1)^2}}{2} \\ 0 \\ 0 \\ 0 \end{pmatrix}$$

Case (b) :

$$f = \begin{pmatrix} \frac{p\_x_2 t\sqrt{(y2-y1)^2+(x2-x1)^2}}{6} \\ 0 \\ \frac{p\_x_2 t\sqrt{(y2-y1)^2+(x2-x1)^2}}{3} \\ 0 \\ 0 \\ 0 \end{pmatrix}$$

Module 3.24: Evaluation of the boundary force matrix in case of a triangular element

Fig. 3.26 Plane elasticity problem of a three-node element under given load and support conditions

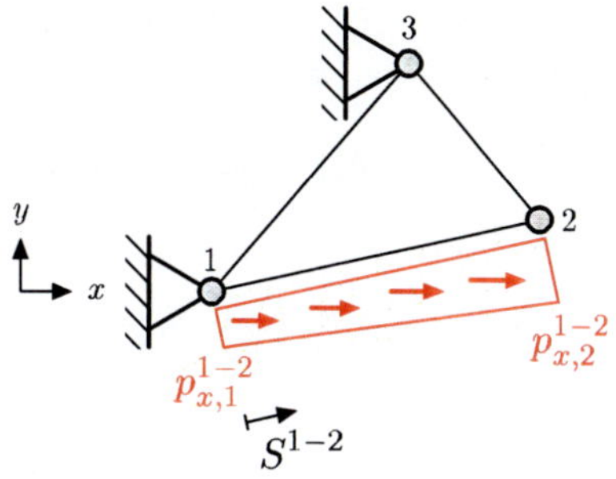

3.14 Example: Plane elasticity problem of a three-node element with distributed load: principal finite element equation

Given is a three-node plane elasticity element as shown in Fig. 3.26. The constant thickness has a value of $t = 0.1$ and the materials parameters are given as $E = 70000$ and $\nu = 0.3$. The lower edge $1-2$ is loaded by a linearly changing pressure (value at node 1: $p_{x,1}^{1-2}$; value at node 2: $p_{x,2}^{1-2}$; both are to be taken as variables), which is acting in the positive x-direction. The nodal coordinates are $(x_1, y_1) = (0, 0)$, $(x_2, y_2) = (5, 0.5)$, and $(x_3, y_3) = (3, 3)$. Derive based on analytical integration the expression for the nodal deformations at node 2.

3.14 Solution

The following Module 3.25 shows the entire wxMaxima code for the nodal deformations at node 2.

```
-->   kill(all);
      load("../Library/my_funs.mac")$
      fpprintprec : 6$ /* float numbers precision (only for printing) */
      ratprint :false$

(%o0) done
```

Definition of Plane Stress Elasticity Problem: Nodal Coordinates "ncoor", Edges defining the Neumann boundaries "edges", Material Properties "mat", Number of Element Nodes "nen", Thickness Function "thickness", freedom tags list "nodtag", freedom values list "nodval", pressures on the edges "pressures"

```
-->   ncoor : [[0,0],[5,0.5],[3,3]]$

      thickness(x,y) := 0.1$

      nodtag : [[1,1],[0,0],[1,1]]$
      nodval : [[0,0],[0,0],[0,0]]$
      edges : [[1,2]]$
      pressures : [[[p_x1,0],[p_x2,0]]]$

      mat : [7.0E4, 0.3]$

      type : "plane stress"$
      nen : 3$
```

Maxima Calculations and Output

```
Ke : k_el_2D_plane_analytical(ncoor, mat, nen, type)$
Fe : transpose(FlatNodePartVector(nodval))$
Fe : Fe + triangular_f_edges_2D_plane_analytical(ncoor, pressures, nen, edges)$
Ke_mod : ModifiedMasterStiffness(nodtag,Ke)$
Fe_mod : ModifiedNodeForces(nodtag,nodval,Ke,Fe)$
Ue : invert(Ke_mod).Fe_mod$

print(" ")$
print("Displacements at Node 2:")$
print(" ", u_2x, "=", float(Ue[3]))$
print(" ", u_2y, "=", float(Ue[4]))$
```

Displacements at Node 2:

$$u_{2x} = [3.14956\ 10^{-5}\ (2.0p_{x2} + p_{x1})]$$

$$u_{2y} = [1.51645\ 10^{-5}\ (2.0p_{x2} + p_{x1})]$$

Module 3.25: Evaluation of the nodal deformations at node 2 for a triangle where lower edge is loaded by a linearly changing pressure

3.15 Example: Plane elasticity problem of a three-node element with distributed load: body force matrix

Given is a three-node plane elasticity element as shown in Fig. 3.27. The constant thickness has a value of t and the materials parameters are given as E and ν. The entire volume is loaded by a constant body force of magnitude f_x, which is acting in the positive x-direction. Derive based on analytical integration the expression for the body force matrix $\boldsymbol{f}_b^{\mathrm{e}}$.

3.15 Solution

The evaluation of the body force matrix $\boldsymbol{f}_b^{\mathrm{e}}$ gives under consideration of $f_y = 0$ and constant thickness t the following expression:

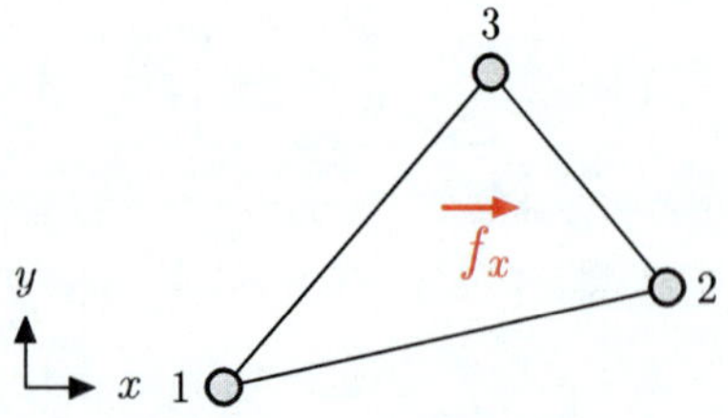

Fig. 3.27 Plane elasticity problem of a three-node element: body force

$$\int_V \boldsymbol{N}\boldsymbol{b}\,\mathrm{d}V = \int_V \begin{bmatrix} N_1 & 0 \\ 0 & N_1 \\ \hline N_2 & 0 \\ 0 & N_2 \\ \hline N_3 & 0 \\ 0 & N_3 \end{bmatrix} \begin{bmatrix} f_x \\ 0 \end{bmatrix} \mathrm{d}V = \int_V \begin{bmatrix} N_1 f_x \\ 0 \\ \hline N_2 f_x \\ 0 \\ \hline N_3 f_x \\ 0 \end{bmatrix} \mathrm{d}V = t \int_A \begin{bmatrix} N_1 f_x \\ 0 \\ \hline N_2 f_x \\ 0 \\ \hline N_3 f_x \\ 0 \end{bmatrix} \mathrm{d}A$$

$$= f_x t \int_{\eta=0}^{1} \int_{\xi=0}^{1-\eta} \begin{bmatrix} \xi \\ 0 \\ \hline \eta \\ 0 \\ \hline 1-\xi-\eta \\ 0 \end{bmatrix} \underbrace{2A}_{J} \, \mathrm{d}\xi \mathrm{d}\eta = 2 f_x t A \int_{\eta=0}^{1} \begin{bmatrix} \frac{1}{2}\xi^2 \\ c \\ \hline \eta\xi \\ c \\ \hline \xi - \frac{1}{2}\xi^2 - \eta\xi \\ c \end{bmatrix}_{\xi=0}^{1-\eta} \mathrm{d}\eta$$

$$= 2 f_x t A \int_{\eta=0}^{1} \begin{bmatrix} \frac{1}{2}\left(1-2\eta+\eta^2\right) \\ c \\ \hline \eta-\eta^2 \\ c \\ \hline \frac{1}{2}\left(1-2\eta+\eta^2\right) \\ c \end{bmatrix} \mathrm{d}\eta = 2 f_x t A \begin{bmatrix} \frac{1}{2}\left(\eta-\eta^2+\frac{1}{3}\eta^3\right) \\ 0 \\ \hline \frac{1}{2}\eta^2 - \frac{1}{3}\eta^3 \\ 0 \\ \hline \frac{1}{2}\left(\eta-\eta^2+\frac{1}{3}\eta^3\right) \\ 0 \end{bmatrix}_0^1$$

$$= 2 f_x t A \begin{bmatrix} \frac{1}{6} \\ 0 \\ \hline \frac{1}{6} \\ 0 \\ \hline \frac{1}{6} \\ 0 \end{bmatrix} = \frac{f_x t A}{3} \begin{bmatrix} 1 \\ 0 \\ \hline 1 \\ 0 \\ \hline 1 \\ 0 \end{bmatrix}. \tag{3.127}$$

The integration can be alternatively performed based on Eq. (3.69):

$$\int_V \boldsymbol{N}\boldsymbol{b}\,\mathrm{d}V = \cdots = f_x t \int_A \begin{bmatrix} \xi \\ 0 \\ \hline \eta \\ 0 \\ \hline \zeta \\ 0 \end{bmatrix} \mathrm{d}x\mathrm{d}y = \frac{f_x t A}{3} \begin{bmatrix} 1 \\ 0 \\ \hline 1 \\ 0 \\ \hline 1 \\ 0 \end{bmatrix}.$$

The following Module 3.26 shows the entire wxMaxima code for the body force matrix.

```
-->   kill(all);
      load("../Library/my_funs.mac")$
      fpprintprec : 6$ /* float numbers precision (only for printing) */
      ratprint : false$

(%o0) done
```

Definition of Vector of Body Forces: Nodal Coordinates "ncoor", Body Forces "body", Number of Element Nodes "nen", Thickness Function "thickness"

```
-->   ncoor : [[x1,y1],[x2,y2],[x3,y3]]$

      thickness(x,y) := t$

      body : [f_x, 0]$

      nen : 3$
```

Maxima Calculations and Output

```
-->   body_forces : body_2D_plane_analytical(ncoor, nen, body)$

      print(" ")$
      print(" Vector of body forces:")$
      print(" f = ", 2*f_x*t*A, body_forces/(f_x*t*jac))$
```

Vector of body forces:

$$f = 2A\, f_x t \begin{pmatrix} \frac{1}{6} \\ 0 \\ \frac{1}{6} \\ 0 \\ \frac{1}{6} \\ 0 \end{pmatrix}$$

Module 3.26: Evaluation of the body force matrix in case of a triangular element

3.16 Example: Plane elasticity problem of a three-node element with body force: principal finite element equation

Given is a three-node plane elasticity element as shown in Fig. 3.28. The constant thickness has a value of $t = 0.1$ and the materials parameters are given as $E = 70000$ and $\nu = 0.3$. The entire volume is loaded by a body force of magnitude f_x, which

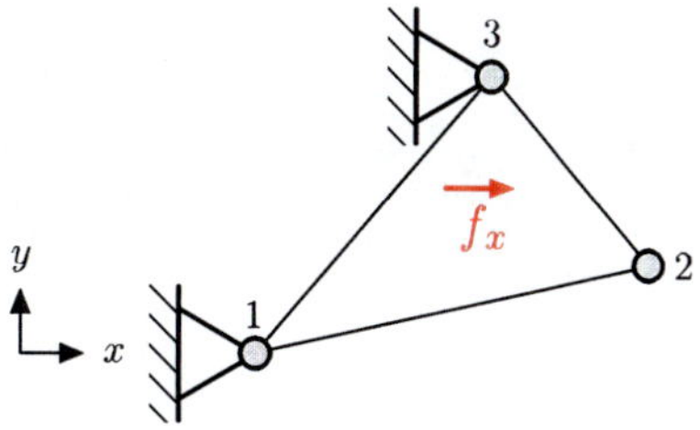

Fig. 3.28 Plane elasticity problem of a three-node element under given body load and support conditions

is acting in the positive x-direction. The nodal coordinates are $(x_1, y_1) = (0, 0)$, $(x_2, y_2) = (5, 0.5)$, and $(x_3, y_3) = (3, 3)$. Derive based on analytical integration the expression for the nodal deformations at node 2.

3.16 Solution

The following Module 3.27 shows the entire wxMaxima code for the nodal deformations at node 2.

```
-->     kill(all);
        load("../Library/my_funs.mac")$
        fpprintprec : 6$ /* float numbers precision (only for printing) */
        ratprint :false$

(% o0)  done
```

Definition of Plane Stress Elasticity Problem: Nodal Coordinates "ncoor", Edges defining the Neumann boundaries "edges", Material Properties "mat", Number of Element Nodes "nen", Thickness Function "thickness", freedom tags list "nodtag", freedom values list "nodval", pressures on the edges "pressures"

```
-->     ncoor : [[0,0],[5,0.5],[3,3]]$

        thickness(x,y) := 0.1$

        nodtag : [[1,1],[0,0],[1,1]]$
        nodval : [[0,0],[0,0],[0,0]]$
        edges : [[1,2]]$
        pressures : [[[0,0],[0,0]]]$
        body : [f_x, 0]$

        mat : [7.0E4, 0.3]$

        type : "plane stress"$
        nen : 3$
```

Maxima Calculations and Output

```
-->     Ke : k_el_2D_plane_analytical(ncoor, mat, nen, type)$
        Fe : transpose(FlatNodePartVector(nodval))$
```

```
Fe : Fe + triangular_f_edges_2D_plane_analytical(ncoor, pressures, nen, edges)$
Fe : Fe + body_2D_plane_analytical(ncoor, nen, body)$
Ke_mod : ModifiedMasterStiffness(nodtag,Ke)$
Fe_mod : ModifiedNodeForces(nodtag,nodval,Ke,Fe)$
Ue : invert(Ke_mod).Fe_mod$

print(" ")$
print("Displacements at Node 2:")$
print(" ", u_2x, "=", float(Ue[3]))$
print(" ", u_2y, "=", float(Ue[4]))$

Displacements at Node 2:
```

$$u_{2x} = [8.46161\ 10^{-5}\ f_x]$$

$$u_{2y} = [4.07411\ 10^{-5}\ f_x]$$

Module 3.27: Evaluation of the nodal deformations at node 2 for a triangle with a body force matrix

3.17 Example: Plate under tensile load based on triangular elements

Given is a two-dimensional member as shown in Fig. 3.29. The left-hand nodes are fixed and the right-hand nodes are loaded by (a) horizontal forces F_0 or (b) a constant pressure $p_{x,0}$. Given are $a = 0.75$, $b = 0.5$ and $\nu = 0.2$ as numbers and F_0 or $p_{x,0}$, t, and E remain variable. Assume consistent units.
Use two triangular plane elasticity elements to:

- Derive the general expression for the stiffness matrix under plane stress condition and for the right-hand side of the finite element equation.

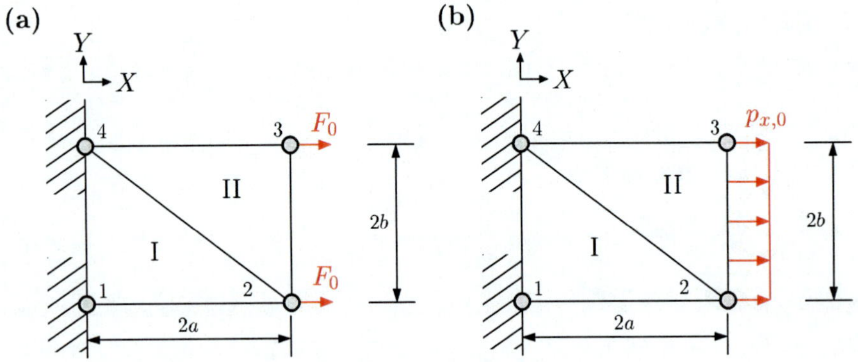

Fig. 3.29 Two-dimensional member under tensile load represented by two triangular elements: **a** concentrated forces and **b** distributed pressure

- Calculate the nodal displacements as a functions of F_0 or $p_{x,0}$, t, and E. Use analytical and numerical 2×2 integration.

3.17 Solution

The following Module 3.28 shows the entire wxMaxima code for the nodal deformations for the concentrated force case.
The following Module 3.29 shows the entire wxMaxima code for the nodal deformations for the distributed pressure case.

```
→       kill(all);
        load("../Library/my_funs.mac")$
        fpprintprec : 6$ /* float numbers precision (only for printing) */
        ratprint :false$

(% o0)  done
```

Definition of Plane Stress Elasticity Problem: Nodal Coordinates "nodxy", Element Connectivity "elenod", Element Material Properties "elemat", Number of Quadrature Points "nqp", Number of Element Nodes "nen", Thickness Function "thickness", freedom tags list "nodtag", freedom values list "nodval", Edges defining the Neumann boundaries "edges", pressures on the edges "pressures", Numerical or Analytical Calculations Switch "numerical"

```
→       a : 0.75$
        b : 0.5$

        nodxy : [[0,-b],[2*a,-b],[2*a,b],[0,b]]$
        elenod : [[1,2,4],[2,3,4]]$
        elemat : [[E, 0.2],[E, 0.2]]$

        numnod : length(nodxy)$
        numele : length(elenod)$

        thickness(x,y) := t$

        nodtag : makelist([0,0],numnod)$
        nodval : makelist([0,0],numnod)$
        nodtag[1] : [1,1]$
        nodtag[4] : [1,1]$
        nodval[2] : [F_0,0]$
        nodval[3] : [F_0,0]$
        edges : [[1,2],[2,3],[3,4],[4,1]]$
        pressures : makelist([[0,0],[0,0]],4)$
        body : [0,0]$

        type : "plane stress"$
        numerical : true $
        nen : 3$
        nqp : 4$
```

Maxima Calculations and Output

```
→       [noddis,Kmod,fmod] : PlanarElementsSolution(nodxy,elenod,elemat,edges,
                              pressures,body,nodtag,nodval,numerical)$

        print(" ")$
        print("=========")$
```

```
print("Modified stiffness matrix (numerical):")$
print(" ", float(Kmod))$

print(" ")$
print("=========")$
print("Modified force matrix (numerical):")$
print(" ", float(fmod))$

print(" ")$
print("=========")$
print("Displacements (numerical):")$
for i : 1 thru length(nodxy) do (
    print("Node",i,":",u_X,"=",float(noddis[i,1]),",",u_Y,"=",float(noddis[i,2]))
)$
```

=========
Modified stiffness matrix (numerical):

$$\begin{bmatrix} 1.0 & 0.0 & 0.0 & 0.0 & 0.0 & 0.0 & 0.0 & 0.0 \\ 0.0 & 1.0 & 0.0 & 0.0 & 0.0 & 0.0 & 0.0 & 0.0 \\ 0.0 & 0.0 & 0.659722Et & 0.0 & -0.3125Et & -0.208333Et & 0.0 & 0.0 \\ 0.0 & 0.0 & 0.0 & 0.920139Et & -0.104167Et & -0.78125Et & 0.0 & 0.0 \\ 0.0 & 0.0 & -0.3125Et & -0.104167Et & 0.659722Et & 0.3125Et & 0.0 & 0.0 \\ 0.0 & 0.0 & -0.208333Et & -0.78125Et & 0.3125Et & 0.920139Et & 0.0 & 0.0 \\ 0.0 & 0.0 & 0.0 & 0.0 & 0.0 & 0.0 & 1.0 & 0.0 \\ 0.0 & 0.0 & 0.0 & 0.0 & 0.0 & 0.0 & 0.0 & 1.0 \end{bmatrix}$$

=========
Modified force matrix (numerical):

$$\begin{bmatrix} 0.0 \\ 0.0 \\ F_0 \\ 0.0 \\ F_0 \\ 0.0 \\ 0.0 \\ 0.0 \end{bmatrix}$$

=========
Displacements (numerical):

Node 1 : $u_X = 0.0, u_Y = 0.0$

Node 2 : $u_X = \dfrac{2.84577 F_0}{Et}, u_Y = \dfrac{0.0570537 F_0}{Et}$

Node 3 : $u_X = \dfrac{3.03249 F_0}{Et}, u_Y = -\dfrac{0.337135 F_0}{Et}$

Node 4 : $u_X = 0.0, u_Y = 0.0$

Module 3.28: Evaluation of the nodal deformations for the concentrated force case

```
→   kill(all);
    load("../Library/my_funs.mac")$
    fpprintprec : 6$ /* float numbers precision (only for printing) */
    ratprint :false$

(% o0)  done
```

Definition of Plane Stress Elasticity Problem: Nodal Coordinates "nodxy", Element Connectivity "elenod", Element Material Properties "elemat", Number of Quadrature Points "nqp", Number of Element Nodes "nen", Thickness Function "thickness", freedom tags list "nodtag", freedom values list "nodval", Edges defining the Neumann boundaries "edges", pressures on the edges "pressures", Numerical or Analytical Calculations Switch "numerical"

```
→   a : 0.75$
    b : 0.5$

    nodxy : [[0,-b],[2*a,-b],[2*a,b],[0,b]]$
    elenod : [[1,2,4],[2,3,4]]$
    elemat : [[E, 0.2],[E, 0.2]]$

    numnod : length(nodxy)$
    numele : length(elenod)$

    thickness(x,y) := t$

    nodtag : makelist([0,0],numnod)$
    nodval : makelist([0,0],numnod)$
    nodtag[1] : [1,1]$
    nodtag[4] : [1,1]$
    edges : [[1,2],[2,3],[3,4],[4,1]]$
    pressures : makelist([[0,0],[0,0]],4)$
    pressures[2] : [[p_x0,0],[p_x0,0]]$
    body : [0,0]$

    type : "plane stress"$
    numerical : true $
    nen : 3$
    nqp : 4$
```

Maxima Calculations and Output

```
→   [noddis,Kmod,fmod] : PlanarElementsSolution(nodxy,elenod,elemat,edges,
                             pressures,body,nodtag,nodval,numerical)$

    print(" ")$
    print("=========")$
    print("Modified stiffness matrix (numerical):")$
    print(" ", float(Kmod))$
```

```
print(" ")$
print("=========")$
print("Modified force matrix (numerical):")$
print(" ", float(fmod))$

print(" ")$
print("=========")$
print("Displacements (numerical):")$
for i : 1 thru length(nodxy) do (
    print("Node",i,":",u_X,"=",float(noddis[i,1]),",",u_Y,"=",float(noddis[i,2]))
)$
```

=========
Modified stiffness matrix (numerical):

$$\begin{bmatrix} 1.0 & 0.0 & 0.0 & 0.0 & 0.0 & 0.0 & 0.0 & 0.0 \\ 0.0 & 1.0 & 0.0 & 0.0 & 0.0 & 0.0 & 0.0 & 0.0 \\ 0.0 & 0.0 & 0.659722Et & 0.0 & -0.3125Et & -0.208333Et & 0.0 & 0.0 \\ 0.0 & 0.0 & 0.0 & 0.920139Et & -0.104167Et & -0.78125Et & 0.0 & 0.0 \\ 0.0 & 0.0 & -0.3125Et & -0.104167Et & 0.659722Et & 0.3125Et & 0.0 & 0.0 \\ 0.0 & 0.0 & -0.208333Et & -0.78125Et & 0.3125Et & 0.920139Et & 0.0 & 0.0 \\ 0.0 & 0.0 & 0.0 & 0.0 & 0.0 & 0.0 & 1.0 & 0.0 \\ 0.0 & 0.0 & 0.0 & 0.0 & 0.0 & 0.0 & 0.0 & 1.0 \end{bmatrix}$$

=========
Modified force matrix (numerical):

$$\begin{bmatrix} 0.0 \\ 0.0 \\ 0.5p_{x0}t \\ 0.0 \\ 0.5p_{x0}t \\ 0.0 \\ 0.0 \\ 0.0 \end{bmatrix}$$

=========
Displacements (numerical):

Node 1 : $u_X = 0.0, u_Y = 0.0$

Node 2 : $U_x = \frac{1.42288p_{x0}}{E}, U_y = \frac{0.0285268p_{x0}}{E}$

Node 3 : $U_x = \frac{1.51624p_{x0}}{E}, U_y = -\frac{0.168568p_{x0}}{E}$

Node 4 : $u_X = 0.0, u_Y = 0.0$

Module 3.29: Evaluation of the nodal deformations for the distributed pressure case

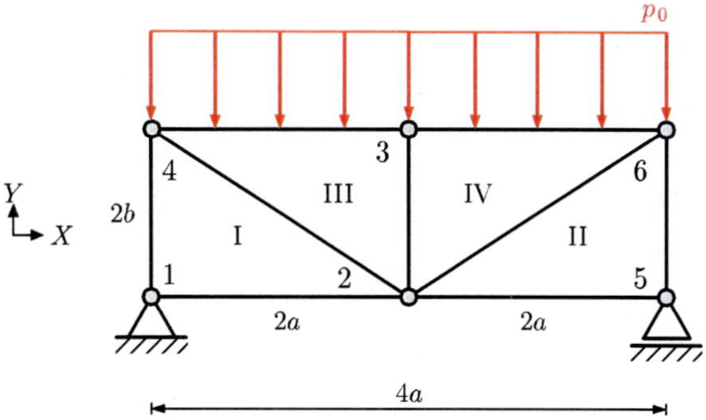

Fig. 3.30 Simply supported beam with constant pressure: modeling approach based on triangular elements (I, ..., IV)

3.18 Advanced Example Based on Triangular Elements: Simply supported beam with distributed load

Given is a finite element representation of simply supported beam as indicated Fig. 3.30. The length of the beam is $4a$ and the rectangular cross section has the dimensions $t \times 2b$. The beam is loaded by a constant pressure p_0 acting in the upper edge of the beam.

Use four plane elasticity triangular elements to calculate the nodal unknowns at node 2 under the assumption of a plane *stress* case and a magnitude of the pressure of $|p_0| = \frac{F_0}{2at}$. Base your solution on analytical integration.

3.18 Solution

The following Module 3.30 shows the entire wxMaxima code for the nodal deformations at node 2.

```
→     kill(all);
      load("../Library/my_funs.mac")$
      fpprintprec : 6$ /* float numbers precision (only for printing) */
      ratprint :false$

(% o0)  done
```

Definition of Plane Stress Elasticity Problem: Nodal Coordinates "nodxy", Element Connectivity "elenod", Element Material Properties "elemat", Number of Quadrature Points "nqp", Number of Element Nodes "nen", Thickness Function "thickness", freedom tags list "nodtag", freedom values list "nodval", Edges defining the Neumann boundaries "edges", pressures on the edges "pressures", Numerical or Analytical Calculations Switch "numerical"

```
→     a : 0.75$
      b : 0.5$
      p_0 : F_0/(2*a*t)$

      nodxy : [[0,-b],[2*a,-b],[2*a,b],[0,b],[4*a,-b],[4*a,b]]$
      elenod : [[1,2,4],[2,5,6],[2,3,4],[2,6,3]]$
      elemat : makelist([E, 0.2],4)$
```

```
numnod : length(nodxy)$
numele : length(elenod)$

thickness(x,y) := t$

nodtag : makelist([0,0],numnod)$
nodval : makelist([0,0],numnod)$
nodtag[1] : [1,1]$
nodtag[5] : [0,1]$
edges : [[6,3,4]]$
pressures : [[[0,-p_0],[0,-p_0],[0,-p_0]]]$
body : [0,0]$

type : "plane stress"$
numerical : false $
nen : 3$
```

Maxima Calculations and Output

```
-->     [noddis,~,~] : PlanarElementsSolution(nodxy,elenod,elemat,edges,pressures,body,
                    nodtag,nodval,numerical)$

        print(" ")$
        print("Displacements at Node 2:")$
        print(" ", u_2X, "=", float(noddis[2][1]))$
        print(" ", u_2Y, "=", float(noddis[2][2]))$
```

Displacements at Node 2:

$$u_{2X} = \frac{1.03077 F_0}{E t}$$

$$u_{2Y} = -\frac{3.6 F_0}{E t}$$

Module 3.30: Evaluation of the nodal deformations at node 2 for the simply supported beam with constant pressure (modeling approach based on triangular elements)

References

1. Akin JE (2005) Finite element analysis with error estimators: an introduction to the FEM and adaptive error analysis for engineering students. Elsevier Butterworth-Heinemann, Burlington
2. Fish J, Belytschko T (2013) A first course in finite elements. Wiley, Chichester
3. MacNeal RH (1994) Finite elements: their design and performance. Marcel Dekker, New York
4. Öchsner A (2016) Continuum damage and fracture mechanics. Springer, Singapore
5. Öchsner A (2020) Computational statics and dynamics - an introduction based on the finite element method. Springer, Singapore

Chapter 4
Classical Plate Elements

4.1 Calculation of the Elemental Stiffness Matrix

A classical plate is defined as a thin structural member, as schematically shown in Fig. 4.1, with a much smaller thickness h than the planar dimensions ($2a$ and $2b$). It can be seen as a two-dimensional extension or generalization of the Euler-Bernoulli beam (see [2]). The following derivations are restricted to some simplifications:

- the thickness h is constant or a function $h = h(x, y)$ but much smaller than the planer dimensions a and b: $\frac{h}{a}$ and $\frac{h}{b} < 0.1$,
- the undeformed plate shape (midplane) is planar,
- the displacement $u_z(x, y)$ is small compared to the thickness dimension h: $u_z < 0.2h$,
- the material is isotropic, homogeneous and linear-elastic according to Hooke's law for a plane stress state ($\sigma_z = \tau_{xz} = \tau_{yz} = 0$), see Eq. (3.1),
- Bernoulli's hypothesis is valid, i.e., a cross-sectional plane stays plane and unwarped in the deformed state. This means that the shear strains γ_{yz} and γ_{xz} due to the distributed shear forces q_x and q_y are neglected,
- external forces act only perpendicular to the xy-plane, the vector of external moments lies within the xy-plane, and
- only rectangular plates are considered.

The external loads, which are considered within this chapter, are single forces F_z, single moments M_x and M_y, area distributed forces $q_z(x, y)$, and area distributed moments $m_x(x, y)$ and $m_y(x, y)$.

The classical theories of plate bending distinguish between shear-rigid and shear-flexible models. The shear rigid-plate, also called the classical or Kirchhoff plate, neglects the shear deformation from the shear forces. This theory corresponds to the classical Euler-Bernoulli beam theory (see [1, 3]). The consideration of the shear deformation leads to the Reissner-Mindlin plate (see Chap. 5) which corresponds to the Timoshenko beam theory (see [1, 3]).

A. Öchsner and R. Makvandi, *Plane Finite Elements for Two-Dimensional Problems*,
https://doi.org/10.1007/978-3-030-89550-1_4

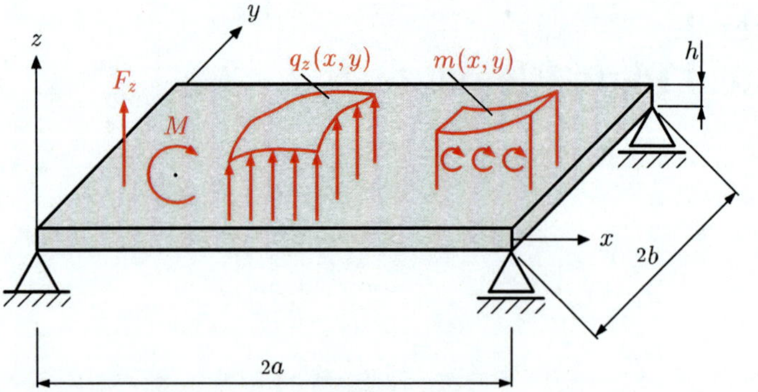

Fig. 4.1 General configuration for a classical plate problem

In addition to the elasticity matrix $\boldsymbol{C}$ in the plane stress formulation of Hooke's law as given in Eq. (3.1), one uses in the context of classical plate theories the so-called plate elasticity matrix $\boldsymbol{D}$ given by [3]

$$\boldsymbol{D} = \frac{h^3}{12}\boldsymbol{C} = \underbrace{\frac{Eh^3}{12(1-\nu^2)}}_{D}\begin{bmatrix} 1 & \nu & 0 \\ \nu & 1 & 0 \\ 0 & 0 & \frac{1-\nu}{2} \end{bmatrix}, \tag{4.1}$$

where $D = \frac{Eh^3}{12(1-\nu^2)}$ is the bending rigidity of the plate.

4.1.1 Elements with Four Nodes

Let us consider in the following the four-node plate element as shown in Fig. 4.2. Important for the following derivations is that the node numbering is counterclockwise. Furthermore, all the following derivations are restricted to the plane stress assumption.

The general expression for the elemental stiffness matrix $\boldsymbol{K}^{\mathrm{e}}$ of a four-node classical plate element with elasticity matrix $\boldsymbol{C}$ can be expressed as

$$\boldsymbol{K}^{\mathrm{e}} = \int_A \underbrace{\left(\boldsymbol{\mathcal{L}}_2\boldsymbol{N}^{\mathrm{T}}\right)^{\mathrm{T}}}_{\boldsymbol{B}} \boldsymbol{D} \underbrace{\left(\boldsymbol{\mathcal{L}}_2\boldsymbol{N}^{\mathrm{T}}\right)}_{\boldsymbol{B}^{\mathrm{T}}} \mathrm{d}A\,, \tag{4.2}$$

where the $\boldsymbol{B}$-matrix contains the second-order derivatives of the interpolation functions N_i $(i = 1, \ldots, 12)$[1]:

[1] Each node has three degrees of freedom and each degree its own interpolation function.

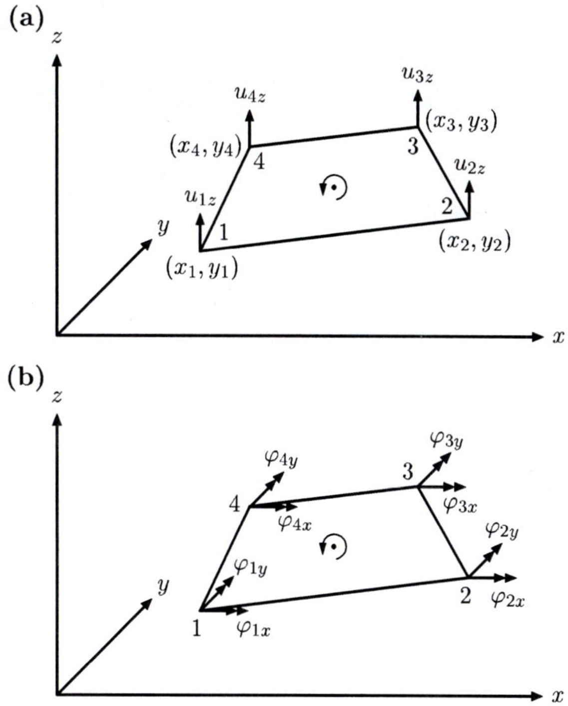

Fig. 4.2 Four-node plate element in the Cartesian space (x, y, z): **a** translational and **b** rotational degrees of freedom

$$\boldsymbol{B} = \left(\boldsymbol{\mathcal{L}}_2 \boldsymbol{N}^{\mathrm{T}}\right)^{\mathrm{T}} = \boldsymbol{N}\boldsymbol{\mathcal{L}}_2^{\mathrm{T}} = \begin{bmatrix} N_1 \\ N_2 \\ \vdots \\ N_n \end{bmatrix} \begin{bmatrix} \frac{\partial^2}{\partial x^2} & \frac{\partial^2}{\partial y^2} & 2\frac{\partial^2}{\partial x \partial y} \end{bmatrix} = \tag{4.3}$$

$$= \begin{bmatrix} \frac{\partial^2 N_1}{\partial x^2} & \frac{\partial^2 N_1}{\partial y^2} & 2\frac{\partial^2 N_1}{\partial x \partial y} \\ \frac{\partial^2 N_2}{\partial x^2} & \frac{\partial^2 N_2}{\partial y^2} & 2\frac{\partial^2 N_2}{\partial x \partial y} \\ \vdots & \vdots & \vdots \\ \frac{\partial^2 N_n}{\partial x^2} & \frac{\partial^2 N_n}{\partial y^2} & 2\frac{\partial^2 N_n}{\partial x \partial y} \end{bmatrix}, \tag{4.4}$$

which is a (12×3)-matrix. The transposed, i.e., $\boldsymbol{B}^{\mathrm{T}} = \boldsymbol{\mathcal{L}}_2 \boldsymbol{N}^{\mathrm{T}}$, is thus a (3×12)-matrix. The triple matrix multiplication gives after integration the stiffness matrix with a dimension of 12×12.

The integration for the element matrix given in Eq. (4.2) is again approximated by numerical integration in commercial finite element packages. To this end, the coordinates (x, y) are transformed to the natural coordinates (ξ, η). In the scope of the coordinate transformation, attention must be paid to the derivatives. The first-order derivatives can be stated under consideration of the chain rule in the following manner:

$$\frac{\partial N(\xi,\eta)}{\partial x} = \frac{\partial N}{\partial \xi}\frac{\partial \xi}{\partial x} + \frac{\partial N}{\partial \eta}\frac{\partial \eta}{\partial x}, \tag{4.5}$$

$$\frac{\partial N(\xi,\eta)}{\partial y} = \frac{\partial N}{\partial \xi}\frac{\partial \xi}{\partial y} + \frac{\partial N}{\partial \eta}\frac{\partial \eta}{\partial y}. \tag{4.6}$$

These results for the first-order derivatives can be used to derive the calculation rules for the second-order derivatives under consideration of the product rule and different coordinate systems:

$$\begin{aligned}
\frac{\partial^2 N(\xi,\eta)}{\partial x^2} &= \frac{\partial}{\partial x}\left(\frac{\partial N(\xi,\eta)}{\partial x}\right) = \frac{\partial}{\partial x}\left(\frac{\partial N}{\partial \xi}\frac{\partial \xi}{\partial x} + \frac{\partial N}{\partial \eta}\frac{\partial \eta}{\partial x}\right), \\
&= \frac{\partial\left(\frac{\partial N}{\partial \xi}\right)}{\partial \xi}\frac{\partial \xi}{\partial x}\frac{\partial \xi}{\partial x} + \frac{\partial\left(\frac{\partial N}{\partial \xi}\right)}{\partial \eta}\frac{\partial \eta}{\partial x}\frac{\partial \xi}{\partial x} + \frac{\partial N}{\partial \xi}\frac{\partial^2 \xi}{\partial x^2} + \\
&+ \frac{\partial\left(\frac{\partial N}{\partial \eta}\right)}{\partial \xi}\frac{\partial \xi}{\partial x}\frac{\partial \eta}{\partial x} + \frac{\partial\left(\frac{\partial N}{\partial \eta}\right)}{\partial \eta}\frac{\partial \eta}{\partial x}\frac{\partial \eta}{\partial x} + \frac{\partial N}{\partial \eta}\frac{\partial^2 \eta}{\partial x^2} \\
&= \frac{\partial^2 N}{\partial \xi^2}\left(\frac{\partial \xi}{\partial x}\right)^2 + \frac{\partial^2 N}{\partial \xi \partial \eta}\frac{\partial \eta}{\partial x}\frac{\partial \xi}{\partial x} + \frac{\partial N}{\partial \xi}\frac{\partial^2 \xi}{\partial x^2} + \\
&+ \frac{\partial^2 N}{\partial \eta \partial \xi}\frac{\partial \xi}{\partial x}\frac{\partial \eta}{\partial x} + \frac{\partial^2 N}{\partial \eta^2}\left(\frac{\partial \eta}{\partial x}\right)^2 + \frac{\partial N}{\partial \eta}\frac{\partial^2 \eta}{\partial x^2},
\end{aligned} \tag{4.7}$$

or finally as:

$$\begin{aligned}
\frac{\partial^2 N(\xi,\eta)}{\partial x^2} &= \frac{\partial^2 N}{\partial \xi^2}\left(\frac{\partial \xi}{\partial x}\right)^2 + \frac{\partial^2 N}{\partial \eta^2}\left(\frac{\partial \eta}{\partial x}\right)^2 + 2\frac{\partial^2 N}{\partial \xi \partial \eta}\frac{\partial \xi}{\partial x}\frac{\partial \eta}{\partial x} + \\
&+ \frac{\partial N}{\partial \xi}\frac{\partial^2 \xi}{\partial x^2} + \frac{\partial N}{\partial \eta}\frac{\partial^2 \eta}{\partial x^2}.
\end{aligned} \tag{4.8}$$

The second-order derivative of an interpolation function with respect to the y-coordinate can be obtained from Eq. (4.8) by substituting x by y. The second-order mixed derivative is obtained in the following way:

$$
\begin{aligned}
\frac{\partial^2 N(\xi,\eta)}{\partial x \partial y} &= \frac{\partial}{\partial y}\left(\frac{\partial N(\xi,\eta)}{\partial x}\right) = \frac{\partial}{\partial y}\left(\frac{\partial N}{\partial \xi}\frac{\partial \xi}{\partial x} + \frac{\partial N}{\partial \eta}\frac{\partial \eta}{\partial x}\right),\\
&= \frac{\partial\left(\frac{\partial N}{\partial \xi}\right)}{\partial y}\frac{\partial \xi}{\partial x} + \frac{\partial N}{\partial \xi}\frac{\partial^2 \xi}{\partial x \partial y} + \frac{\partial\left(\frac{\partial N}{\partial \eta}\right)}{\partial y}\frac{\partial \eta}{\partial x} + \frac{\partial N}{\partial \eta}\frac{\partial^2 \eta}{\partial x \partial y}\\
&= \frac{\partial^2 N}{\partial \xi^2}\frac{\partial \xi}{\partial y}\frac{\partial \xi}{\partial x} + \frac{\partial^2 N}{\partial \xi \partial \eta}\frac{\partial \eta}{\partial y}\frac{\partial \xi}{\partial x} + \frac{\partial N}{\partial \xi}\frac{\partial^2 \xi}{\partial x \partial y} +\\
&\quad + \frac{\partial^2 N}{\partial \eta \partial \xi}\frac{\partial \xi}{\partial y}\frac{\partial \eta}{\partial x} + \frac{\partial^2 N}{\partial \eta^2}\frac{\partial \eta}{\partial y}\frac{\partial \eta}{\partial x} + \frac{\partial N}{\partial \eta}\frac{\partial^2 \eta}{\partial x \partial y},
\end{aligned}
\tag{4.9}
$$

or finally as:

$$
\begin{aligned}
\frac{\partial^2 N(\xi,\eta)}{\partial x \partial y} &= \frac{\partial^2 N}{\partial \xi^2}\frac{\partial \xi}{\partial x}\frac{\partial \xi}{\partial y} + \frac{\partial^2 N}{\partial \eta^2}\frac{\partial \eta}{\partial x}\frac{\partial \eta}{\partial y} + \frac{\partial^2 N}{\partial \xi \partial \eta}\left(\frac{\partial \xi}{\partial x}\frac{\partial \eta}{\partial y} + \frac{\partial \xi}{\partial y}\frac{\partial \eta}{\partial x}\right) +\\
&\quad + \frac{\partial N}{\partial \xi}\frac{\partial^2 \xi}{\partial x \partial y} + \frac{\partial N}{\partial \eta}\frac{\partial^2 \eta}{\partial x \partial y}.
\end{aligned}
\tag{4.10}
$$

The relationships for the second-order derivatives can be combined in matrix notation in the following way:

$$
\begin{bmatrix} \frac{\partial^2 N}{\partial x^2} \\ \frac{\partial^2 N}{\partial y^2} \\ \frac{\partial^2 N}{\partial x \partial y} \end{bmatrix}
\begin{bmatrix}
\left(\frac{\partial \xi}{\partial x}\right)^2 & \left(\frac{\partial \eta}{\partial x}\right)^2 & 2\frac{\partial \xi}{\partial x}\frac{\partial \eta}{\partial x} \\
\left(\frac{\partial \xi}{\partial y}\right)^2 & \left(\frac{\partial \eta}{\partial y}\right)^2 & 2\frac{\partial \xi}{\partial y}\frac{\partial \eta}{\partial y} \\
\frac{\partial \xi}{\partial x}\frac{\partial \xi}{\partial y} & \frac{\partial \eta}{\partial x}\frac{\partial \eta}{\partial y} & \frac{\partial \xi}{\partial x}\frac{\partial \eta}{\partial y} + \frac{\partial \xi}{\partial y}\frac{\partial \eta}{\partial x}
\end{bmatrix}
\begin{bmatrix} \frac{\partial^2 N}{\partial \xi^2} \\ \frac{\partial^2 N}{\partial \eta^2} \\ \frac{\partial^2 N}{\partial \xi \partial \eta} \end{bmatrix}
+
\begin{bmatrix}
\frac{\partial^2 \xi}{\partial x^2} & \frac{\partial^2 \eta}{\partial x^2} \\
\frac{\partial^2 \xi}{\partial y^2} & \frac{\partial^2 \eta}{\partial y^2} \\
\frac{\partial^2 \xi}{\partial x \partial y} & \frac{\partial^2 \eta}{\partial x \partial y}
\end{bmatrix}
\begin{bmatrix} \frac{\partial N}{\partial \xi} \\ \frac{\partial N}{\partial \eta} \end{bmatrix}.
\tag{4.11}
$$

Furthermore, the coordinate transformation requires that $\mathrm{d}A = \mathrm{d}x\mathrm{d}y \rightarrow \mathrm{d}V' = J\mathrm{d}\xi\mathrm{d}\eta$, where J is the Jacobian:

$$
J = \left|\frac{\partial(x,y)}{\partial(\xi,\eta)}\right| = \frac{\partial x}{\partial \xi} \times \frac{\partial y}{\partial \eta} - \frac{\partial x}{\partial \eta} \times \frac{\partial y}{\partial \xi}.
\tag{4.12}
$$

In order to proceed, we need to have a closer look on the interpolation functions N_i. Under the assumption of a fourth-order polynomial for the displacement field $u_z^e(\xi,\eta)$ in the parametric ξ-η space and four nodes per element, the following twelve interpolation functions can be easily derived for a regular four-node plate element,[2] see [3] for details:

[2] All internal angles equal to 90°, outer dimensions $2a \times 2b$ in the x- and y-direction.

$$N_{1u} = N_1 = -\frac{(\eta - 1)(\xi - 1)\left(\xi^2 + \eta^2 + \xi + \eta - 2\right)}{8}, \tag{4.13}$$

$$N_{1\varphi_x} = N_2 = -\frac{b(\eta + 1)(\eta - 1)^2(\xi - 1)}{8}, \tag{4.14}$$

$$N_{1\varphi_y} = N_3 = \frac{a(\xi + 1)(\xi - 1)^2(\eta - 1)}{8}, \tag{4.15}$$

$$N_{2u} = N_4 = \frac{(\eta - 1)(\xi + 1)\left(\xi^2 + \eta^2 - \xi + \eta - 2\right)}{8}, \tag{4.16}$$

$$N_{2\varphi_x} = N_5 = \frac{b(\eta + 1)(\eta - 1)^2(\xi + 1)}{8}, \tag{4.17}$$

$$N_{2\varphi_y} = N_6 = \frac{a(\xi - 1)(\xi + 1)^2(\eta - 1)}{8}, \tag{4.18}$$

$$N_{3u} = N_7 = -\frac{(\eta + 1)(\xi + 1)\left(\xi^2 + \eta^2 - \xi - \eta - 2\right)}{8}, \tag{4.19}$$

$$N_{3\varphi_x} = N_8 = \frac{b(\eta - 1)(\eta + 1)^2(\xi + 1)}{8}, \tag{4.20}$$

$$N_{3\varphi_y} = N_9 = -\frac{a(\xi - 1)(\xi + 1)^2(\eta + 1)}{8}, \tag{4.21}$$

$$N_{4u} = N_{10} = \frac{(\eta + 1)(\xi - 1)\left(\xi^2 + \eta^2 + \xi - \eta - 2\right)}{8}, \tag{4.22}$$

$$N_{4\varphi_x} = N_{11} = -\frac{b(\eta - 1)(\eta + 1)^2(\xi - 1)}{8}, \tag{4.23}$$

$$N_{4\varphi_y} = N_{12} = -\frac{a(\xi + 1)(\xi - 1)^2(\eta + 1)}{8}. \tag{4.24}$$

It should be noted here that these twelve interpolation functions can be written in compact form ($i = 1, \ldots, 4$) as follows [4]:

$$N_{iu} = \frac{1}{8} \times (1 + \xi\xi_i)(1 + \eta\eta_i)(2 + \xi\xi_i + \eta\eta_i - \xi^2 - \eta^2), \tag{4.25}$$

$$N_{i\varphi_x} = \frac{b}{8} \times \eta_i(1 + \xi\xi_i)(\eta\eta_i - 1)(1 + \eta\eta_i)^2, \tag{4.26}$$

$$N_{i\varphi_y} = -\frac{a}{8} \times \xi_i(\xi\xi_i - 1)(1 + \eta\eta_i)(1 + \xi\xi_i)^2. \tag{4.27}$$

The corresponding interpolation functions for node 1 are shown in Fig. 4.3. Comparing with the characteristics of the cubic interpolation functions for a beam element (see [3]), the same characteristics can be identified: The interpolation function for the displacement takes at its own node a value of one and is at all other nodes equal to zero. The interpolation functions for the rotations are at all nodes equal to zero but the slope takes absolute values of either one or zero.

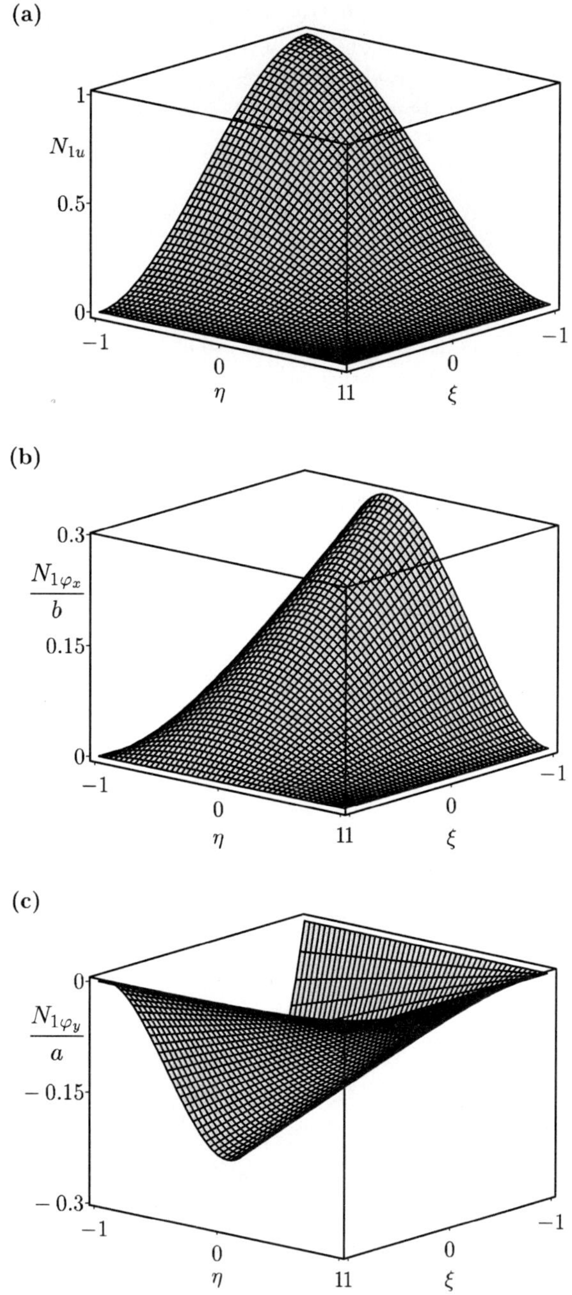

Fig. 4.3 Interpolation functions at node 1 ($\xi = -1, \eta = -1$) of a rectangular plate element (three DOF per node): **a** displacement; **b** and **c** rotations

The evaluation of the second-order derivatives with respect to the Cartesian coordinates (x, y) as indicated in Eq. (4.11) requires the calculation of the first- and second-order derivatives with respect to the natural coordinates (ξ, η). These derivatives can be obtained from Eqs. (4.25)–(4.27) in the following way:

$$\frac{\partial N_{iu}}{\partial \xi} = -\tfrac{1}{8}(\eta\eta_i + 1)(\eta^2\xi_i - \eta\eta_i\xi_i + 3\xi^2\xi_i - 2\xi\xi_i^2 + 2\xi - 3\xi_i)\,, \tag{4.28}$$

$$\frac{\partial N_{iu}}{\partial \eta} = -\tfrac{1}{8}(\xi\xi_i + 1)(3\eta^2\eta_i - 2\eta\eta_i^2 + \eta_i\xi^2 - \eta_i\xi\xi_i + 2\eta - 3\eta_i)\,, \tag{4.29}$$

$$\frac{\partial^2 N_{iu}}{\partial \xi^2} = -\tfrac{1}{4}(\eta\eta_i + 1)(3\xi\xi_i - \xi_i^2 + 1)\,, \tag{4.30}$$

$$\frac{\partial^2 N_{iu}}{\partial \eta^2} = -\tfrac{1}{4}(\xi\xi_i + 1)(3\eta\eta_i - \eta_i^2 + 1)\,, \tag{4.31}$$

$$\begin{aligned}\frac{\partial^2 N_{iu}}{\partial \xi\partial\eta} = -\tfrac{1}{8}\,\big(&\xi_i(3\eta^2\eta_i - 2\eta\eta_i^2 + \eta_i\xi^2 - \eta_i\xi\xi_i + 2\eta - 3\eta_i) \\ &+ (\xi\xi_i + 1)(2\eta_i\xi - \eta_i\xi_i)\big)\,.\end{aligned} \tag{4.32}$$

$$\frac{\partial N_{i\varphi_x}}{\partial \xi} = \tfrac{1}{8}b\eta_i\xi_i(\eta\eta_i - 1)(\eta\eta_i + 1)^2\,, \tag{4.33}$$

$$\frac{\partial N_{i\varphi_x}}{\partial \eta} = \tfrac{1}{8}b\eta_1^2(\xi\xi_i + 1)(\eta\eta_i + 1)(3\eta\eta_i - 1)\,, \tag{4.34}$$

$$\frac{\partial^2 N_{i\varphi_x}}{\partial \xi^2} = 0\,, \tag{4.35}$$

$$\frac{\partial^2 N_{i\varphi_x}}{\partial \eta^2} = \tfrac{1}{4}b\eta_i^3(\xi\xi_i + 1)(3\eta\eta_i + 1)\,, \tag{4.36}$$

$$\frac{\partial^2 N_{i\varphi_x}}{\partial \xi\partial\eta} = \tfrac{1}{8}b\eta_i^2\xi_1(\eta\eta_i + 1)(3\eta\eta_i - 1)\,. \tag{4.37}$$

$$\frac{\partial N_{i\varphi_y}}{\partial \xi} = -\tfrac{1}{8}a\xi_i^2(\eta\eta_i + 1)(\xi\xi_i + 1)(3\xi\xi_i - 1)\,, \tag{4.38}$$

$$\frac{\partial N_{i\varphi_y}}{\partial \eta} = -\tfrac{1}{8}a\xi_i(\xi\xi_i - 1)\eta_i(\xi\xi_i + 1)^2\,, \tag{4.39}$$

$$\frac{\partial^2 N_{i\varphi_y}}{\partial \xi^2} = -\tfrac{1}{4}a\xi_i^3(\eta\eta_i + 1)(3\xi\xi_i + 1)\,, \tag{4.40}$$

$$\frac{\partial^2 N_{i\varphi_y}}{\partial \eta^2} = 0\,, \tag{4.41}$$

$$\frac{\partial^2 N_{i\varphi_y}}{\partial \xi\partial\eta} = -\tfrac{1}{8}a\xi_i^2\eta_i(\xi\xi_i + 1)(3\xi\xi_1 - 1)\,. \tag{4.42}$$

To calculate the geometrical derivatives, let us assume the same interpolation for the global x- and y-coordinate as for a four-node plane elasticity element (see Sect. 3.1.1)[3]:

$$x(\xi,\eta) = \overline{N}_1(\xi,\eta)\times x_1 + \overline{N}_2(\xi,\eta)\times x_2 + \overline{N}_3(\xi,\eta)\times x_3 + \overline{N}_4(\xi,\eta)\times x_4\,, \tag{4.43}$$

$$y(\xi,\eta) = \overline{N}_1(\xi,\eta)\times y_1 + \overline{N}_2(\xi,\eta)\times y_2 + \overline{N}_3(\xi,\eta)\times y_3 + \overline{N}_4(\xi,\eta)\times y_4\,, \tag{4.44}$$

where the *linear* shape functions $\overline{N}_i$ are given by Eqs. (3.12)–(3.15) and the *global* coordinates of the nodes $1,\ldots,4$ can be used for $x_1,\ldots,x_4$ and $y_1,\ldots,y_4$.

Thus, the geometrical derivatives can easily be obtained as:

$$\frac{\partial x}{\partial \xi} = \frac{1}{4}\Big((-1+\eta)x_1 + (1-\eta)x_2 + (1+\eta)x_3 + (-1-\eta)x_4\Big)\,, \tag{4.45}$$

$$\frac{\partial y}{\partial \xi} = \frac{1}{4}\Big((-1+\eta)y_1 + (1-\eta)y_2 + (1+\eta)y_3 + (-1-\eta)y_4\Big)\,, \tag{4.46}$$

$$\frac{\partial x}{\partial \eta} = \frac{1}{4}\Big((-1+\xi)x_1 + (-1-\xi)x_2 + (1+\xi)x_3 + (1-\xi)x_4\Big)\,, \tag{4.47}$$

$$\frac{\partial y}{\partial \eta} = \frac{1}{4}\Big((-1+\xi)y_1 + (-1-\xi)y_2 + (1+\xi)y_3 + (1-\xi)y_4\Big)\,. \tag{4.48}$$

The evaluation of Eq. (4.11) requires, however, the geometrical derivatives of the natural coordinates (ξ,η) with respect to the physical coordinates (x,y). These relations can be easily obtained from Eqs. (4.45)–(4.48) under consideration of the following relationships [3]:

$$\frac{\partial \xi}{\partial x} = +\frac{1}{\frac{\partial x}{\partial \xi}\frac{\partial y}{\partial \eta} - \frac{\partial x}{\partial \eta}\frac{\partial y}{\partial \xi}} \times \frac{\partial y}{\partial \eta}\,, \tag{4.49}$$

$$\frac{\partial \xi}{\partial y} = -\frac{1}{\frac{\partial x}{\partial \xi}\frac{\partial y}{\partial \eta} - \frac{\partial x}{\partial \eta}\frac{\partial y}{\partial \xi}} \times \frac{\partial x}{\partial \eta}\,, \tag{4.50}$$

$$\frac{\partial \eta}{\partial x} = -\frac{1}{\frac{\partial x}{\partial \xi}\frac{\partial y}{\partial \eta} - \frac{\partial x}{\partial \eta}\frac{\partial y}{\partial \xi}} \times \frac{\partial y}{\partial \xi}\,, \tag{4.51}$$

$$\frac{\partial \eta}{\partial y} = +\frac{1}{\frac{\partial x}{\partial \xi}\frac{\partial y}{\partial \eta} - \frac{\partial x}{\partial \eta}\frac{\partial y}{\partial \xi}} \times \frac{\partial x}{\partial \xi}\,, \tag{4.52}$$

whereas the second-order derivatives follow as:

---

[3] Now we have the case $\deg(\overline{N}_i) < \deg(N_i)$ and a subparametric element formulation is obtained.

$$\frac{\partial^2 \xi}{\partial x^2} = \frac{\partial}{\partial x}\left(+\frac{1}{\frac{\partial x}{\partial \xi}\frac{\partial y}{\partial \eta} - \frac{\partial x}{\partial \eta}\frac{\partial y}{\partial \xi}} \times \frac{\partial y}{\partial \eta}\right), \tag{4.53}$$

$$\frac{\partial^2 \xi}{\partial y^2} = \frac{\partial}{\partial y}\left(-\frac{1}{\frac{\partial x}{\partial \xi}\frac{\partial y}{\partial \eta} - \frac{\partial x}{\partial \eta}\frac{\partial y}{\partial \xi}} \times \frac{\partial x}{\partial \eta}\right), \tag{4.54}$$

$$\frac{\partial^2 \xi}{\partial x \partial y} = \frac{\partial}{\partial y}\left(+\frac{1}{\frac{\partial x}{\partial \xi}\frac{\partial y}{\partial \eta} - \frac{\partial x}{\partial \eta}\frac{\partial y}{\partial \xi}} \times \frac{\partial y}{\partial \eta}\right), \tag{4.55}$$

$$\frac{\partial^2 \eta}{\partial x^2} = \frac{\partial}{\partial x}\left(-\frac{1}{\frac{\partial x}{\partial \xi}\frac{\partial y}{\partial \eta} - \frac{\partial x}{\partial \eta}\frac{\partial y}{\partial \xi}} \times \frac{\partial y}{\partial \xi}\right), \tag{4.56}$$

$$\frac{\partial^2 \eta}{\partial y^2} = \frac{\partial}{\partial y}\left(+\frac{1}{\frac{\partial x}{\partial \xi}\frac{\partial y}{\partial \eta} - \frac{\partial x}{\partial \eta}\frac{\partial y}{\partial \xi}} \times \frac{\partial x}{\partial \xi}\right), \tag{4.57}$$

$$\frac{\partial^2 \eta}{\partial x \partial y} = \frac{\partial}{\partial y}\left(-\frac{1}{\frac{\partial x}{\partial \xi}\frac{\partial y}{\partial \eta} - \frac{\partial x}{\partial \eta}\frac{\partial y}{\partial \xi}} \times \frac{\partial y}{\partial \xi}\right), \tag{4.58}$$

where, for example, the following rule must be considered: $\frac{\partial}{\partial x}(f(\xi,\eta)) = \frac{\partial f}{\partial \xi}\frac{\partial \xi}{\partial x} + \frac{\partial f}{\partial \eta}\frac{\partial \eta}{\partial x}$.

Based on the derived equations, the triple matrix product $\boldsymbol{B}\boldsymbol{D}\boldsymbol{B}^{\mathrm{T}}$ (see Eq. (4.2)) can be numerically calculated to obtain the stiffness matrix.

The integration in Eq. (4.2) is performed in commercial finite element packages based on Gauss-Legendre quadrature. For the domain integrals, one can refer to Eq. (3.30) and Table 3.1.

Let us summarize here the major steps which are required to calculate the elemental stiffness matrix for a regular four-node plate element.

❶ Introduce an elemental coordinate system (x, y).
❷ Express the coordinates (x_i, y_i) of the corner nodes i $(i = 1, \ldots, 4)$ in this elemental coordinate system.
❸ Calculate the partial derivatives of the Cartesian (x, y) coordinates with respect to the natural (ξ, η) coordinates, see Eqs. (4.45)–(4.48):

$$\frac{\partial x}{\partial \xi} = x_\xi = \frac{1}{4}\Big((-1+\eta)x_1 + (1-\eta)x_2 + (1+\eta)x_3 + (-1-\eta)x_4\Big),$$

$$\vdots$$

$$\frac{\partial y}{\partial \eta} = y_\eta = \frac{1}{4}\Big((-1+\xi)y_1 + (-1-\xi)y_2 + (1+\xi)y_3 + (1-\xi)y_4\Big).$$

❹ Calculate the partial derivatives of the natural (ξ, η) coordinates with respect to the Cartesian (x, y) coordinates. First-order derivatives according to Eqs. (4.49)–(4.52):

$$\frac{\partial \xi}{\partial x} = +\frac{1}{\frac{\partial x}{\partial \xi}\frac{\partial y}{\partial \eta} - \frac{\partial x}{\partial \eta}\frac{\partial y}{\partial \xi}} \times \frac{\partial y}{\partial \eta}, \quad \frac{\partial \xi}{\partial y} = -\frac{1}{\frac{\partial x}{\partial \xi}\frac{\partial y}{\partial \eta} - \frac{\partial x}{\partial \eta}\frac{\partial y}{\partial \xi}} \times \frac{\partial x}{\partial \eta},$$

$$\frac{\partial \eta}{\partial x} = -\frac{1}{\frac{\partial x}{\partial \xi}\frac{\partial y}{\partial \eta} - \frac{\partial x}{\partial \eta}\frac{\partial y}{\partial \xi}} \times \frac{\partial y}{\partial \xi}, \quad \frac{\partial \eta}{\partial y} = +\frac{1}{\frac{\partial x}{\partial \xi}\frac{\partial y}{\partial \eta} - \frac{\partial x}{\partial \eta}\frac{\partial y}{\partial \xi}} \times \frac{\partial x}{\partial \xi}.$$

Second-order derivatives according to Eqs. (4.53)–(4.58):

$$\frac{\partial^2 \xi}{\partial x^2} = \frac{\partial}{\partial x}\left(+\frac{1}{\frac{\partial x}{\partial \xi}\frac{\partial y}{\partial \eta} - \frac{\partial x}{\partial \eta}\frac{\partial y}{\partial \xi}} \times \frac{\partial y}{\partial \eta}\right),$$

$$\vdots$$

$$\frac{\partial^2 \eta}{\partial x \partial y} = \frac{\partial}{\partial y}\left(-\frac{1}{\frac{\partial x}{\partial \xi}\frac{\partial y}{\partial \eta} - \frac{\partial x}{\partial \eta}\frac{\partial y}{\partial \xi}} \times \frac{\partial y}{\partial \xi}\right).$$

❺ Calculate the $\boldsymbol{B}$-matrix and its transposed, see Eq. (4.4):

$$\boldsymbol{B}^{\mathrm{T}} = \begin{bmatrix} \frac{\partial^2 N_1}{\partial x^2} & \frac{\partial^2 N_2}{\partial x^2} & \cdots & \frac{\partial^2 N_n}{\partial x^2} \\ \frac{\partial^2 N_1}{\partial y^2} & \frac{\partial^2 N_2}{\partial y^2} & \cdots & \frac{\partial^2 N_n}{\partial y^2} \\ 2\frac{\partial^2 N_1}{\partial x \partial y} & 2\frac{\partial^2 N_2}{\partial x \partial y} & \cdots & 2\frac{\partial^2 N_n}{\partial x \partial y} \end{bmatrix},$$

where the second-order partial derivatives of the interpolation functions are given in Eq. (4.11).

❻ Calculate the triple matrix product $\boldsymbol{B}\boldsymbol{D}\boldsymbol{B}^{\mathrm{T}}$, where the plate elasticity matrix $\boldsymbol{D}$ is given by Eq. (4.1).

❼ Perform the numerical integration, for example, based on a 2 × 2 integration rule:

$$\int_A (\boldsymbol{BDB}^{\mathrm{T}})\mathrm{d}A = \boldsymbol{BDB}^{\mathrm{T}} J \times 1\Big|_{\left(-\frac{1}{\sqrt{3}},-\frac{1}{\sqrt{3}}\right)}$$
$$+ \boldsymbol{BDB}^{\mathrm{T}} J \times 1\Big|_{\left(\frac{1}{\sqrt{3}},-\frac{1}{\sqrt{3}}\right)} + \boldsymbol{BDB}^{\mathrm{T}} J \times 1\Big|_{\left(\frac{1}{\sqrt{3}},\frac{1}{\sqrt{3}}\right)}$$
$$+ \boldsymbol{BDB}^{\mathrm{T}} J \times 1\Big|_{\left(-\frac{1}{\sqrt{3}},\frac{1}{\sqrt{3}}\right)}.$$

❽ $\boldsymbol{K}^{\mathrm{e}}$ obtained.

In the case of a distorted four-node plate element, the above eight steps must be modified since the geometrical derivatives are no more equal to simple constant expression, i.e. $\frac{\partial \eta}{\partial y} \neq \frac{1}{b}$ and $\frac{\partial \xi}{\partial x} \neq \frac{1}{a}$. This directly affects the rotational fields, which must be written as

$$\varphi_x^{\mathrm{e}}(\xi, \eta) = \frac{\partial u_z^{\mathrm{e}}}{\partial y} = \frac{\partial u_z^{\mathrm{e}}(\xi, \eta)}{\partial \eta}\frac{\partial \eta}{\partial y}. \tag{4.59}$$

and

$$\varphi_y^{\mathrm{e}}(\xi,\eta) = -\frac{\partial u_z^{\mathrm{e}}}{\partial x} = -\frac{\partial u_z^{\mathrm{e}}(\xi,\eta)}{\partial \xi}\frac{\partial \xi}{\partial x}, \tag{4.60}$$

where the geometrical derivatives must be calculated according to Eqs. (4.49) and (4.49). As a consequence, the shape function given in Eqs. (4.13)–(4.24), or more precisely the shape functions for the rotational degrees of freedom, must be adjusted as follows:

$$N_{1u} = N_1 = -\frac{(\eta-1)(\xi-1)\left(\xi^2+\eta^2+\xi+\eta-2\right)}{8}, \tag{4.61}$$

$$N_{1\varphi_x} = N_2 = -\left(\frac{\partial \eta}{\partial y}\right)^{-1} \times \frac{(\eta+1)(\eta-1)^2(\xi-1)}{8}, \tag{4.62}$$

$$N_{1\varphi_y} = N_3 = \left(\frac{\partial \xi}{\partial x}\right)^{-1} \times \frac{(\xi+1)(\xi-1)^2(\eta-1)}{8}, \tag{4.63}$$

$$N_{2u} = N_4 = \frac{(\eta-1)(\xi+1)\left(\xi^2+\eta^2-\xi+\eta-2\right)}{8}, \tag{4.64}$$

$$N_{2\varphi_x} = N_5 = \left(\frac{\partial \eta}{\partial y}\right)^{-1} \times \frac{(\eta+1)(\eta-1)^2(\xi+1)}{8}, \tag{4.65}$$

$$N_{2\varphi_y} = N_6 = \left(\frac{\partial \xi}{\partial x}\right)^{-1} \times \frac{(\xi-1)(\xi+1)^2(\eta-1)}{8}, \tag{4.66}$$

$$N_{3u} = N_7 = -\frac{(\eta+1)(\xi+1)\left(\xi^2+\eta^2-\xi-\eta-2\right)}{8}, \tag{4.67}$$

$$N_{3\varphi_x} = N_8 = \left(\frac{\partial \eta}{\partial y}\right)^{-1} \times \frac{(\eta-1)(\eta+1)^2(\xi+1)}{8}, \tag{4.68}$$

$$N_{3\varphi_y} = N_9 = -\left(\frac{\partial \xi}{\partial x}\right)^{-1} \times \frac{(\xi-1)(\xi+1)^2(\eta+1)}{8}, \tag{4.69}$$

$$N_{4u} = N_{10} = \frac{(\eta+1)(\xi-1)\left(\xi^2+\eta^2+\xi-\eta-2\right)}{8}, \tag{4.70}$$

$$N_{4\varphi_x} = N_{11} = -\left(\frac{\partial \eta}{\partial y}\right)^{-1} \times \frac{(\eta-1)(\eta+1)^2(\xi-1)}{8}, \tag{4.71}$$

$$N_{4\varphi_y} = N_{12} = -\left(\frac{\partial \xi}{\partial x}\right)^{-1} \times \frac{(\xi+1)(\xi-1)^2(\eta+1)}{8}. \tag{4.72}$$

Thus, the first- and second-order derivatives of the shape functions for the rotational degrees of freedom must be accordingly adjusted.

4.1.2 Maxima Modules for Stiffness Matrices

4.1.3 Maxima Examples to Calculate the Stiffness Matrix

4.1 Example: Stiffness matrix for a two-dimensional rectangular plate element

Given is a regular two-dimensional classical plate element as shown in Fig. 4.4.

Derive the general expression for the elemental stiffness matrix under the plane stress assumption. Furthermore, it can be assumed that the thickness h is constant. Use first analytical integration to obtain the stiffness matrix and then compare the results with a 2×2 Gauss-Legendre integration rule.

4.1 Solution

The following Module 4.1 shows the entire wxMaxima code for the derivation of the stiffness matrix based on analytical integration.

The following Module 4.2 shows the entire wxMaxima code for the derivation of the stiffness matrix based on 2×2 numerical integration.

4.2 Example: Distorted two-dimensional plate element

Given is a distorted two-dimensional classical plate element as shown in Fig. 4.5. The

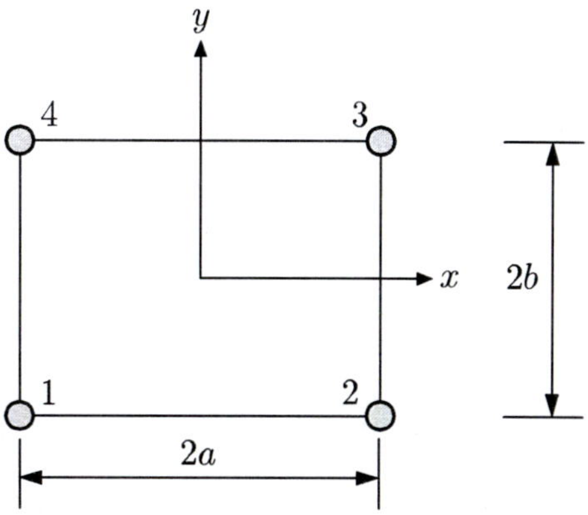

Fig. 4.4 Regular two-dimensional classical plate element

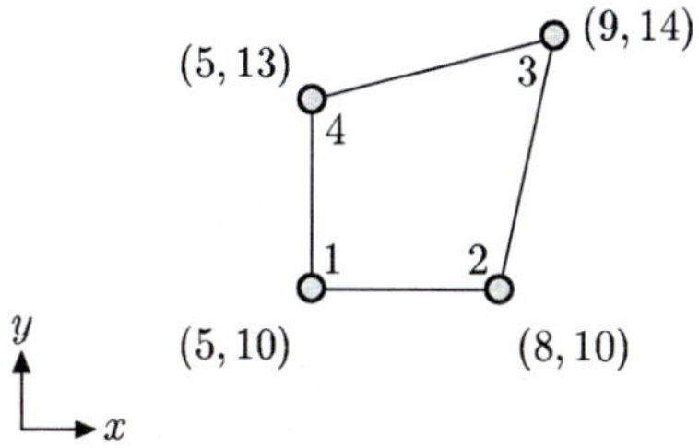

Fig. 4.5 Distorted two-dimensional plate element

```
→    kill(all);
     load("../Library/my_funs.mac")$
     ratprint :false$

(% o0)  done
```

Definition of Plane Stress Stiffness: Nodal Coordinates "ncoor", Material Properties "mat", Thickness Function "thickness", Number of Element Nodes "nen"

```
→    ncoor : [[-a,-b],[a,-b],[a,b],[-a,b]]$

     mat : [E, v]$

     thickness(x,y) := h$

     nen : 4$
```

Maxima Calculations and Output

```
→    Ke : k_el_2D_Kirschhoff_plate_analytical(ncoor, mat, nen)$

     print(" ")$
     print("Element stiffness matrix:")$
     print(" ")$
     print(K_el = matrix([K_el[1,1],K_el[1,2]],[K_el[2,1],K_el[2,2]]))$
     print(" ")$
     coeff : mat[1]*(th^3)/(12*(1-mat[2]^2))$
     print(D, "=", coeff)$

     for i : 1 thru 2 do (
       for j : 1 thru 2 do (
         Ke_l[i,j] : zeromatrix(6,6)
       )
     )$

     for i : 1 thru 6 do (
       for j : 1 thru 6 do (
         Ke_l[1,1][i,j] : factor(ratsimp(float(Ke[i,j]/coeff))),
         Ke_l[1,2][i,j] : factor(ratsimp(float(Ke[i,j+6]/coeff))),
         Ke_l[2,1][i,j] : factor(ratsimp(float(Ke[i+6,j]/coeff))),
         Ke_l[2,2][i,j] : factor(ratsimp(float(Ke[i+6,j+6]/coeff)))
       )
     )$

     for i : 1 thru 2 do (
       for j : 1 thru 2 do (
         print(" "),
         print(K_el[i,j], "="),
         print(D, Ke_l[i,j])
       )
     )$
```

Element stiffness matrix:

$$K_{el} = \begin{bmatrix} K_{el1,1} & K_{el1,2} \\ K_{el2,1} & K_{el2,2} \end{bmatrix}$$

$$D = \frac{E\,h^3}{12\left(1 - v^2\right)}$$

$$K_{el1,1} = D \begin{bmatrix} -\frac{2a^2b^2v-10b^4-7a^2b^2-10a^4}{10a^3b^3} & \frac{4b^2v+b^2+10a^2}{10ab^2} & \cdots & \frac{a^2v-10b^2-a^2}{10a^2b} \\ \frac{4b^2v+b^2+10a^2}{10ab^2} & -\frac{4\left(b^2v-b^2-5a^2\right)}{15ab} & \cdots & 0 \\ -\frac{4a^2v+10b^2+a^2}{10a^2b} & -v & \cdots & \frac{a^2v+10b^2-a^2}{15ab} \\ \frac{2a^2b^2v-10b^4-7a^2b^2+5a^4}{10a^3b^3} & -\frac{4b^2v+b^2-5a^2}{10ab^2} & \cdots & \frac{4a^2v+10b^2+a^2}{10a^2b} \\ -\frac{4b^2v+b^2-5a^2}{10ab^2} & \frac{2\left(2b^2v-2b^2+5a^2\right)}{15ab} & \cdots & v \\ \frac{a^2v-10b^2-a^2}{10a^2b} & 0 & \cdots & -\frac{4\left(a^2v-5b^2-a^2\right)}{15ab} \end{bmatrix}$$

$$K_{el1,2} = D \begin{bmatrix} -\frac{2a^2b^2v+5b^4-7a^2b^2+5a^4}{10a^3b^3} & \frac{b^2v-b^2+5a^2}{10ab^2} & \cdots & \frac{4a^2v-5b^2+a^2}{10a^2b} \\ -\frac{b^2v-b^2+5a^2}{10ab^2} & -\frac{b^2v-b^2-5a^2}{15ab} & \cdots & 0 \\ \frac{a^2v+5b^2-a^2}{10a^2b} & 0 & \cdots & \frac{2\left(2a^2v+5b^2-2a^2\right)}{15ab} \\ \frac{2a^2b^2v+5b^4-7a^2b^2-10a^4}{10a^3b^3} & -\frac{b^2v-b^2-10a^2}{10ab^2} & \cdots & \frac{a^2v+5b^2-a^2}{10a^2b} \\ \frac{b^2v-b^2-10a^2}{10ab^2} & \frac{b^2v-b^2+10a^2}{15ab} & \cdots & 0 \\ -\frac{4a^2v-5b^2+a^2}{10a^2b} & 0 & \cdots & -\frac{a^2v-5b^2-a^2}{15ab} \end{bmatrix}$$

$$K_{el2,1} = D \begin{bmatrix} -\frac{2a^2b^2v+5b^4-7a^2b^2+5a^4}{10a^3b^3} & -\frac{b^2v-b^2+5a^2}{10ab^2} & \cdots & -\frac{4a^2v-5b^2+a^2}{10a^2b} \\ \frac{b^2v-b^2+5a^2}{10ab^2} & -\frac{b^2v-b^2-5a^2}{15ab} & \cdots & 0 \\ -\frac{a^2v+5b^2-a^2}{10a^2b} & 0 & \cdots & \frac{2\left(2a^2v+5b^2-2a^2\right)}{15ab} \\ \frac{2a^2b^2v+5b^4-7a^2b^2-10a^4}{10a^3b^3} & \frac{b^2v-b^2-10a^2}{10ab^2} & \cdots & -\frac{a^2v+5b^2-a^2}{10a^2b} \\ -\frac{b^2v-b^2-10a^2}{10ab^2} & \frac{b^2v-b^2+10a^2}{15ab} & \cdots & 0 \\ \frac{4a^2v-5b^2+a^2}{10a^2b} & 0 & \cdots & -\frac{a^2v-5b^2-a^2}{15ab} \end{bmatrix}$$

$$K_{el2,2} = D \begin{bmatrix} -\frac{2a^2b^2v-10b^4-7a^2b^2-10a^4}{10a^3b^3} & -\frac{4b^2v+b^2+10a^2}{10ab^2} & \cdots & -\frac{a^2v-10b^2-a^2}{10a^2b} \\ -\frac{4b^2v+b^2+10a^2}{10ab^2} & -\frac{4\left(b^2v-b^2-5a^2\right)}{15ab} & \cdots & 0 \\ \frac{4a^2v+10b^2+a^2}{10a^2b} & -v & \cdots & \frac{a^2v+10b^2-a^2}{15ab} \\ \frac{2a^2b^2v-10b^4-7a^2b^2+5a^4}{10a^3b^3} & \frac{4b^2v+b^2-5a^2}{10ab^2} & \cdots & -\frac{4a^2v+10b^2+a^2}{10a^2b} \\ \frac{4b^2v+b^2-5a^2}{10ab^2} & \frac{2\left(2b^2v-2b^2+5a^2\right)}{15ab} & \cdots & v \\ -\frac{a^2v-10b^2-a^2}{10a^2b} & 0 & \cdots & -\frac{4\left(a^2v-5b^2-a^2\right)}{15ab} \end{bmatrix}$$

Module 4.1: Stiffness matrix for a regular ($2a \times 2b \times h$) two-dimensional classical plate element based on analytical integration

```
→   kill(all);
    load("../Library/my_funs.mac")$
    ratprint :false$

(% o0)  done
```

Definition of Plane Stress Stiffness: Nodal Coordinates "ncoor", Material Properties "mat", Thickness Function "thickness", Number of Element Nodes "nen", Number of Quadrature Points "nqp"

```
→   ncoor : [[-a,-b],[a,-b],[a,b],[-a,b]]$

    mat : [E, v]$

    thickness(x,y) := h$

    nqp : 4$
    nen : 4$
```

Maxima Calculations and Output

```
→   Ke : k_el_2D_Kirschhoff_plate_numerical(ncoor, mat, nen, nqp)$

    print(" ")$
    print("Element stiffness matrix:")$
    print(" ")$
    print(K_el = matrix([K_el[1,1],K_el[1,2]],[K_el[2,1],K_el[2,2]]))$
    print(" ")$
    coeff : mat[1]*(th^3)/(12*(1-mat[2]^2))$
    print(D, "=", coeff)$

    for i : 1 thru 2 do (
      for j : 1 thru 2 do (
        Ke_l[i,j] : zeromatrix(6,6)
      )
    )$

    for i : 1 thru 6 do (
      for j : 1 thru 6 do (
        Ke_l[1,1][i,j] : factor(ratsimp(float(Ke[i,j]/coeff))),
        Ke_l[1,2][i,j] : factor(ratsimp(float(Ke[i,j+6]/coeff))),
        Ke_l[2,1][i,j] : factor(ratsimp(float(Ke[i+6,j]/coeff))),
        Ke_l[2,2][i,j] : factor(ratsimp(float(Ke[i+6,j+6]/coeff)))
      )
    )$

    for i : 1 thru 2 do (
      for j : 1 thru 2 do (
        print(" "),
        print(K_el[i,j], "="),
        print(D, Ke_l[i,j])
      )
    )$
```

Element stiffness matrix:

$$K_{el} = \begin{bmatrix} K_{el1,1} & K_{el1,2} \\ K_{el2,1} & K_{el2,2} \end{bmatrix}$$

$$D = \frac{E\,h^3}{12\,(1 - v^2)}$$

$$K_{el1,1} = D \begin{bmatrix} \frac{2b^4+a^2b^2+2a^4}{2a^3b^3} & \frac{b^2v+2a^2}{2ab^2} & \cdots & -\frac{b}{a^2} \\ \frac{b^2v+2a^2}{2ab^2} & -\frac{b^2v-b^2-8a^2}{6ab} & \cdots & 0 \\ -\frac{a^2v+2b^2}{2a^2b} & -v & \cdots & \frac{a^2v+4b^2-a^2}{6ab} \\ -\frac{(b^2+a^2)(2b^2-a^2)}{2a^3b^3} & -\frac{b^2v-a^2}{2ab^2} & \cdots & \frac{a^2v+2b^2}{2a^2b} \\ -\frac{b^2v-a^2}{2ab^2} & \frac{b^2v-b^2+4a^2}{6ab} & \cdots & v \\ -\frac{b}{a^2} & 0 & \cdots & -\frac{a^2v-8b^2-a^2}{6ab} \end{bmatrix}$$

$$K_{el1,2} = D \begin{bmatrix} -\frac{b^4-a^2b^2+a^4}{2a^3b^3} & \frac{a}{2b^2} & \cdots & \frac{a^2v-b^2}{2a^2b} \\ -\frac{a}{2b^2} & -\frac{b^2v-b^2-2a^2}{6ab} & \cdots & 0 \\ \frac{b}{2a^2} & 0 & \cdots & \frac{a^2v+4b^2-a^2}{6ab} \\ \frac{(b^2-2a^2)(b^2+a^2)}{2a^3b^3} & \frac{a}{b^2} & \cdots & \frac{b}{2a^2} \\ -\frac{a}{b^2} & \frac{b^2v-b^2+4a^2}{6ab} & \cdots & 0 \\ -\frac{a^2v-b^2}{2a^2b} & 0 & \cdots & -\frac{a^2v-2b^2-a^2}{6ab} \end{bmatrix}$$

$$K_{el2,1} = D \begin{bmatrix} -\frac{b^4-a^2b^2+a^4}{2a^3b^3} & -\frac{a}{2b^2} & \cdots & -\frac{a^2v-b^2}{2a^2b} \\ \frac{a}{2b^2} & -\frac{b^2v-b^2-2a^2}{6ab} & \cdots & 0 \\ -\frac{b}{2a^2} & 0 & \cdots & \frac{a^2v+4b^2-a^2}{6ab} \\ \frac{(b^2-2a^2)(b^2+a^2)}{2a^3b^3} & -\frac{a}{b^2} & \cdots & -\frac{b}{2a^2} \\ \frac{a}{b^2} & \frac{b^2v-b^2+4a^2}{6ab} & \cdots & 0 \\ \frac{a^2v-b^2}{2a^2b} & 0 & \cdots & -\frac{a^2v-2b^2-a^2}{6ab} \end{bmatrix}$$

$$K_{el2,2} = D \begin{bmatrix} \frac{2b^4+a^2b^2+2a^4}{2a^3b^3} & -\frac{b^2v+2a^2}{2ab^2} & \cdots & \frac{b}{a^2} \\ -\frac{b^2v+2a^2}{2ab^2} & -\frac{b^2v-b^2-8a^2}{6ab} & \cdots & 0 \\ \frac{a^2v+2b^2}{2a^2b} & -v & \cdots & \frac{a^2v+4b^2-a^2}{6ab} \\ -\frac{(b^2+a^2)(2b^2-a^2)}{2a^3b^3} & \frac{b^2v-a^2}{2ab^2} & \cdots & -\frac{a^2v+2b^2}{2a^2b} \\ \frac{b^2v-a^2}{2ab^2} & \frac{b^2v-b^2+4a^2}{6ab} & \cdots & v \\ \frac{b}{a^2} & 0 & \cdots & -\frac{a^2v-8b^2-a^2}{6ab} \end{bmatrix}$$

Module 4.2: Stiffness matrix for a regular ($2a \times 2b \times h$) two-dimensional classical plate element based on 2×2 numerical integration

Table 4.1 Comparison of the results for some selected elements of the elemental stiffness matrix depending on the integration method

Method	K_{11}	K_{12}	K_{13}
Analytical	1.11306	0.834252	−0.834252
2 × 2	0.994145	0.732575	−0.732575
3 × 3	1.11253	0.833658	−0.833658
4 × 4	1.11304	0.834236	−0.834236
5 × 5	1.11306	0.834252	−0.834252
6 × 6	1.11306	0.834252	−0.834252

constant thickness has a value of $h = 0.1$ and the materials parameters are given as $E = 70000$ and $\nu = 0.3$. Derive the numerical expression for the elemental stiffness matrix under the plane stress assumption. Use analytical, 2×2 Gauss-Legendre, and 6×6 Gauss-Legendre integration. Compare the accuracy of the integration approach for a few selected elements of the stiffness matrix.

4.2 Solution

The following Module 4.3 shows the entire wxMaxima code for the derivation of the stiffness matrix based on 6×6 numerical integration.

The different results for some selected elements of the elemental stiffness matrix depending on the integration method are shown in Table 4.1.

4.2 Elemental and Global Principal Finite Element Equation

4.2.1 Elements with Four Nodes

The principal finite element equation for a single classical four-node plate element (based on the derivations for the stiffness matrix given in Sect. 4.1.1) can be stated as follows (see [3])

```
→     kill(all);
      load("../Library/my_funs.mac")$
      ratprint :false$ fpprintprec : 6$

(% o0)  done
```

Definition of Plane Stress Stiffness: Nodal Coordinates "ncoor", Material Properties "mat", Thickness Function "thickness", Number of Element Nodes "nen", Number of Quadrature Points "nqp"

```
→     ncoor : [[5,10],[8,10],[9,14],[5,13]]$

      mat : [70000, 3/10]$

      thickness(x,y) := 1/10$

      nqp : 36$
      nen : 4$
```

Maxima Calculations and Output

```
→     Ke : k_el_2D_Kirschhoff_plate_numerical(ncoor, mat, nen, nqp)$

      print(" ")$
      print("Element stiffness matrix:")$
      print(" ")$
      print(K_el = matrix([K_el[1,1],K_el[1,2]],[K_el[2,1],K_el[2,2]]))$
      print(" ")$
      coeff : mat[1]*(th^3)/(12*(1-mat[2]^2))$
      print(D, "=", coeff)$

      for i : 1 thru 2 do (
        for j : 1 thru 2 do (
          Ke_l[i,j] : zeromatrix(6,6)
        )
      )$

      for i : 1 thru 6 do (
        for j : 1 thru 6 do (
          Ke_l[1,1][i,j] : factor(ratsimp(float(Ke[i,j]/coeff))),
          Ke_l[1,2][i,j] : factor(ratsimp(float(Ke[i,j+6]/coeff))),
          Ke_l[2,1][i,j] : factor(ratsimp(float(Ke[i+6,j]/coeff))),
          Ke_l[2,2][i,j] : factor(ratsimp(float(Ke[i+6,j+6]/coeff)))
        )
      )$

      for i : 1 thru 2 do (
        for j : 1 thru 2 do (
          print(" "),
          print(K_el[i,j], "="),
          print(D, Ke_l[i,j])
        )
      )$
```

Element stiffness matrix:

$$K_{el} = \begin{bmatrix} K_{el1,1} & K_{el1,2} \\ K_{el2,1} & K_{el2,2} \end{bmatrix}$$

$$D = \frac{250}{39}$$

$$K_{el1,1} = D \begin{bmatrix} 1.11306 & 0.834252 & -0.834252 & -0.499758 & 0.218306 & -0.794501 \\ 0.834252 & 1.47103 & -0.327629 & 6.4530710^{-4} & 0.368451 & -0.044531 \\ -0.834252 & -0.327629 & 1.47103 & 0.737144 & -0.0012856 & 0.726714 \\ -0.499758 & 6.4530710^{-4} & 0.737144 & 1.03031 & 0.72291 & 0.884571 \\ 0.218306 & 0.368451 & -0.0012856 & 0.72291 & 1.65676 & 0.454744 \\ -0.794501 & -0.044531 & 0.726714 & 0.884571 & 0.454744 & 2.28211 \end{bmatrix}$$

$$K_{el1,2} = D \begin{bmatrix} -0.11354 & 0.113111 & -0.113111 & -0.499758 & 0.794501 & -0.218306 \\ -0.0977531 & 0.134879 & 0.093017 & -0.737144 & 0.726714 & -0.0012856 \\ 0.0977531 & 0.093017 & 0.134879 & -6.4530710^{-4} & -0.044531 & 0.368451 \\ -0.428407 & 0.643941 & -0.035045 & -0.102146 & 0.528784 & 0.339794 \\ -0.601422 & 0.616315 & -0.15145 & -0.339794 & 0.709906 & 0.0475313 \\ 0.438715 & -0.193859 & 0.619175 & -0.528784 & 0.133424 & 0.709906 \end{bmatrix}$$

$$K_{el2,1} = D \begin{bmatrix} -0.11354 & -0.0977531 & 0.0977531 & -0.428407 & -0.601422 & 0.438715 \\ 0.113111 & 0.134879 & 0.093017 & 0.643941 & 0.616315 & -0.193859 \\ -0.113111 & 0.093017 & 0.134879 & -0.035045 & -0.15145 & 0.619175 \\ -0.499758 & -0.737144 & -6.4530710^{-4} & -0.102146 & -0.339794 & -0.528784 \\ 0.794501 & 0.726714 & -0.044531 & 0.528784 & 0.709906 & 0.133424 \\ -0.218306 & -0.0012856 & 0.368451 & 0.339794 & 0.0475313 & 0.709906 \end{bmatrix}$$

$$K_{el2,2} = D \begin{bmatrix} 0.970354 & -0.792097 & 0.792097 & -0.428407 & -0.438715 & 0.601422 \\ -0.792097 & 1.3106 & -0.288269 & 0.035045 & 0.619175 & -0.15145 \\ 0.792097 & -0.288269 & 1.3106 & -0.643941 & -0.193859 & 0.616315 \\ -0.428407 & 0.035045 & -0.643941 & 1.03031 & -0.884571 & -0.72291 \\ -0.438715 & 0.619175 & -0.193859 & -0.884571 & 2.28211 & 0.454744 \\ 0.601422 & -0.15145 & 0.616315 & -0.72291 & 0.454744 & 1.65676 \end{bmatrix}$$

Module 4.3: Stiffness matrix for a distorted two-dimensional classical plate element based on 6×6 numerical integration

$$
\begin{bmatrix}
K_{11} & K_{12} & K_{13} & \dots & K_{110} & K_{111} & K_{112} \\
K_{21} & K_{22} & K_{23} & \dots & K_{210} & K_{211} & K_{212} \\
K_{31} & K_{32} & K_{33} & \dots & K_{310} & K_{311} & K_{312} \\
\vdots & \vdots & \vdots & \vdots & \vdots & \vdots & \vdots \\
K_{101} & K_{102} & K_{103} & \dots & K_{1010} & K_{1011} & K_{1012} \\
K_{111} & K_{112} & K_{113} & \dots & K_{1110} & K_{1111} & K_{1112} \\
K_{121} & K_{122} & K_{123} & \dots & K_{1210} & K_{1211} & K_{1212}
\end{bmatrix}
\begin{bmatrix}
u_{1z} \\ \varphi_{1x} \\ \varphi_{1y} \\ \hline u_{2z} \\ \varphi_{2x} \\ \varphi_{2y} \\ \hline u_{3z} \\ \varphi_{3x} \\ \varphi_{3y} \\ \hline u_{4z} \\ \varphi_{4x} \\ \varphi_{4y}
\end{bmatrix}
=
\begin{bmatrix}
F_{1z} \\ M_{1x} \\ M_{1y} \\ \hline F_{2z} \\ M_{2x} \\ M_{2y} \\ \hline F_{3z} \\ M_{3x} \\ M_{3y} \\ \hline F_{4z} \\ M_{4x} \\ M_{4y}
\end{bmatrix}
+ \int_A
\begin{bmatrix}
N_{1u} \\ N_{1\varphi_x} \\ N_{1\varphi_y} \\ \hline N_{2u} \\ N_{2\varphi_x} \\ N_{2\varphi_y} \\ \hline N_{3u} \\ N_{3\varphi_x} \\ N_{3\varphi_y} \\ \hline N_{4u} \\ N_{4\varphi_x} \\ N_{4\varphi_y}
\end{bmatrix}
q_z \, \mathrm{d}A \,, \tag{4.73}
$$

or in abbreviated form

$$
\boldsymbol{K}^{\mathrm{e}} \boldsymbol{u}_{\mathrm{p}}^{\mathrm{e}} = \boldsymbol{f}_{\boldsymbol{t}}^{\mathrm{e}} + \boldsymbol{f}_{\boldsymbol{b}}^{\mathrm{e}} \,, \tag{4.74}
$$

where $\boldsymbol{K}^{\mathrm{e}}$ is the elemental stiffness matrix (the calculation is to be performed according to the scheme given on Sect. 4.1.1), $\boldsymbol{u}_{\mathrm{p}}^{\mathrm{e}}$ is the column matrix of nodal unknowns, $\boldsymbol{f}_{\boldsymbol{t}}^{\mathrm{e}}$ is the boundary force column matrix, and $\boldsymbol{f}_{\boldsymbol{b}}^{\mathrm{e}}$ is body force column matrix. The twelve interpolation functions N_i are given in Eqs. (4.13)–(4.24) for a regular element or in Eqs. (4.61)–(4.72) for a distorted element.

Let us have now a closer look on the body force column matrix $\boldsymbol{f}_{\boldsymbol{b}}^{\mathrm{e}}$. This integral can be written under consideration of the natural coordinates (ξ, η) as follows:

$$
\boldsymbol{f}_{\boldsymbol{b}}^{\mathrm{e}} = \int_A
\begin{bmatrix}
N_{1u} \\ N_{1\varphi_x} \\ N_{1\varphi_y} \\ N_{2u} \\ N_{2\varphi_x} \\ N_{2\varphi_y} \\ N_{3u} \\ N_{3\varphi_x} \\ N_{3\varphi_y} \\ N_{4u} \\ N_{4\varphi_x} \\ N_{4\varphi_y}
\end{bmatrix}
q_z(x, y) \mathrm{d}A = \int_{-1}^{1} \int_{-1}^{1}
\begin{bmatrix}
N_{1u}(\xi, \eta) \\ N_{1\varphi_x}(\xi, \eta) \\ N_{1\varphi_y}(\xi, \eta) \\ N_{2u}(\xi, \eta) \\ N_{2\varphi_x}(\xi, \eta) \\ N_{2\varphi_y}(\xi, \eta) \\ N_{3u}(\xi, \eta) \\ N_{3\varphi_x}(\xi, \eta) \\ N_{3\varphi_y}(\xi, \eta) \\ N_{4u}(\xi, \eta) \\ N_{4\varphi_x}(\xi, \eta) \\ N_{4\varphi_y}(\xi, \eta)
\end{bmatrix}
q_z(\xi, \eta) J(\xi, \eta) \mathrm{d}\xi \mathrm{d}\eta \,, \tag{4.75}
$$

where the Jacobian $J = J(\xi, \eta)$ can be derived from Eq. 4.12.

Table 4.2 Post-processing of nodal values for a classical four-node plate element. The distributions are given as being dependent on the nodal values $\boldsymbol{u}_{\mathrm{p}}^{\mathrm{e}}$ as a function of the natural coordinates $-1 \leq \xi \leq +1$ and $-1 \leq \eta \leq +1$

Displacement field
$u_z^{\mathrm{e}}(\xi,\eta) = \boldsymbol{N}^{\mathrm{T}}(\xi,\eta)\boldsymbol{u}_{\mathrm{p}}^{\mathrm{e}}$
Rotational fields
$\varphi_x^{\mathrm{e}}(\xi,\eta) = \frac{\partial u_z^{\mathrm{e}}(\xi,\eta)}{\partial \eta}\frac{\partial \eta}{\partial y}$
$\varphi_y^{\mathrm{e}}(\xi,\eta) = -\frac{\partial u_z^{\mathrm{e}}(\xi,\eta)}{\partial \xi}\frac{\partial \xi}{\partial x}$
Strain field
$\boldsymbol{\varepsilon}^{\mathrm{e}}(\xi,\eta) = -z\boldsymbol{\mathcal{L}}_2 u_z^{\mathrm{e}}(\xi,\eta)$
Stress field
$\boldsymbol{\sigma}^{\mathrm{e}}(\xi,\eta) = -z\boldsymbol{C}\boldsymbol{\mathcal{L}}_2 u_z^{\mathrm{e}}(\xi,\eta)$

Once the nodal deformations are known, e.g. based on $\boldsymbol{u}_{\mathrm{p}}^{\mathrm{e}} = (\boldsymbol{K}^{\mathrm{e}})^{-1}\boldsymbol{f}^{\mathrm{e}}$, further quantities can be calculated based on this result (the so-called post-processing, see Table 4.2).

Let us now have a closer look on the field equations provided in Table 4.2. The displacement field can be written as

$$u_z^{\mathrm{e}}(\xi,\eta) = \left[N_{1u}\ N_{1\varphi_x}\ N_{1\varphi_y} \middle| \ldots \middle| N_{4u}\ N_{4\varphi_x}\ N_{4\varphi_y}\right] \begin{bmatrix} u_{1z} \\ \varphi_{1x} \\ \varphi_{1y} \\ \hline \vdots \\ \hline u_{4z} \\ \varphi_{4x} \\ \varphi_{4y} \end{bmatrix}, \tag{4.76}$$

where the interpolation functions $N_i(\xi,\eta)$ are given in Eqs. (4.13)–(4.24) for a regular element or in Eqs. (4.61)–(4.72) for a distorted element. The rotational fields can be written as:

$$\varphi_x^{\mathrm{e}}(\xi,\eta) = \left[\frac{\partial N_{1u}}{\partial \eta}\ \frac{\partial N_{1\varphi_x}}{\partial \eta}\ \frac{\partial N_{1\varphi_y}}{\partial \eta} \middle| \ldots \middle| \frac{\partial N_{4u}}{\partial \eta}\ \frac{\partial N_{4\varphi_x}}{\partial \eta}\ \frac{\partial N_{4\varphi_y}}{\partial \eta}\right] \begin{bmatrix} u_{1z} \\ \varphi_{1x} \\ \varphi_{1y} \\ \hline \vdots \\ \hline u_{4z} \\ \varphi_{4x} \\ \varphi_{4y} \end{bmatrix} \frac{\partial \eta}{\partial y}, \tag{4.77}$$

$$\varphi_y^{\mathrm{e}}(\xi,\eta) = -\left[\frac{\partial N_{1u}}{\partial \xi}\;\frac{\partial N_{1\varphi_x}}{\partial \xi}\;\frac{\partial N_{1\varphi_y}}{\partial \xi}\middle|\ldots\middle|\frac{\partial N_{4u}}{\partial \xi}\;\frac{\partial N_{4\varphi_x}}{\partial \xi}\;\frac{\partial N_{4\varphi_y}}{\partial \xi}\right]\begin{bmatrix} u_{1z} \\ \varphi_{1x} \\ \varphi_{1y} \\ \hline \vdots \\ \hline u_{4z} \\ \varphi_{4x} \\ \varphi_{4y} \end{bmatrix}\frac{\partial \xi}{\partial x}. \tag{4.78}$$

For the calculation of the strain and stress fields, the second-order derivatives of the interpolation functions with respect to the Cartesian coordinates are required, see the $\mathcal{L}_2$-operator. This can be done based on the relations provided in Eq. (4.11).

4.2.2 *Maxima Modules for the Finite Element Equation*

4.2.3 *Maxima Examples for the Principal Finite Element Equation*

4.3 Example: One-element example of a cantilever plate

Given is a cantilever classical plate element as indicated in Fig. 4.6. The side lengths are equal to $2a$. The plate is loaded by two single forces $\frac{1}{2}F_0$ acting at the right-hand nodes of the element. The material is described based on the engineering constants Young's modulus E and Poisson's ratio ν.

Use a single plate element in the following to model the problem and to calculate the nodal unknowns based on analytical and numerical integration. Compare your results with the analytical solutions for a cantilever Euler-Bernoulli and a Timoshenko beam.

Fig. 4.6 Cantilever square plate element

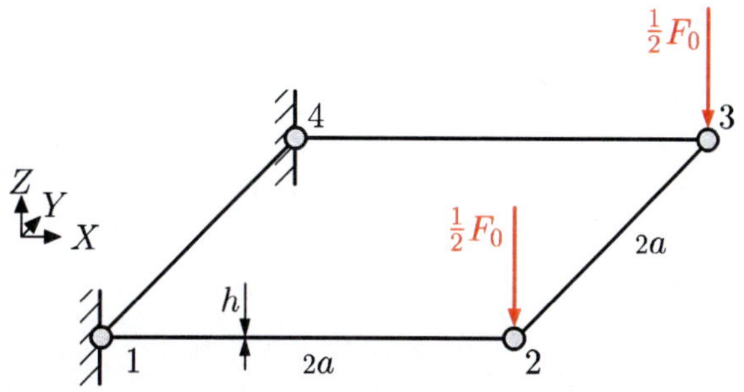

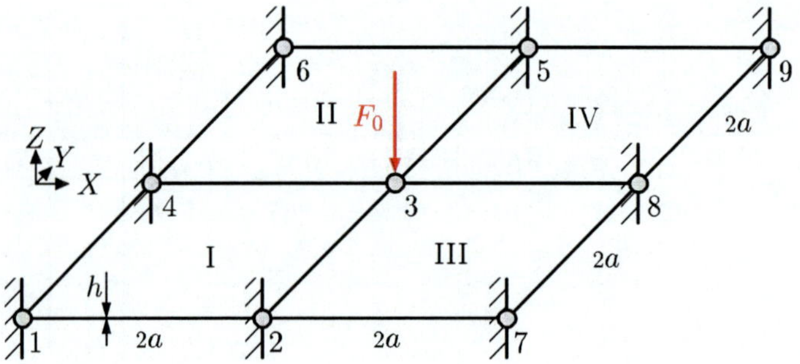

Fig. 4.7 Plate problem with four edges fixed

4.3 Solution
The following Module 4.4 shows the entire wxMaxima code for the calculation of the nodal unknowns based on analytical integration.

The following Module 4.5 shows the entire wxMaxima code for the calculation of the nodal unknowns based on numerical integration.

The analytical solution is obtained as $u_{Z,\max} = -\frac{16Fa^2}{Eh^3}$ for the Euler-Bernoulli beam and as $u_{Z,\max} = -\frac{16Fa^2}{Eh^3} - \frac{6F}{5hG}$ for the Timoshenko beam. It should be noted here that the Euler-Bernoulli solution is obtained as a special case from u_{2Z} for $\nu \rightarrow 0$.

4.4 Example: Four-element example of a plate fixed at all four edges

Given is a classical plate which is fixed at all four sides, see Fig. 4.7. The side lengths are equal to $4a$. The plate is loaded by a single forces F_0 acting in the middle of the plate. The material is described based on the engineering constants Young's modulus E and Poisson's ratio ν.

Use four classical plate elements (each $2a \times 2a \times h$) in the following to model the problem and to calculate the nodal unknowns in the middle of the plate based on analytical and numerical integration.

Use the analytical solution provided in [5] to benchmark the finite element solution.

4.4 Solution
The following Module 4.6 shows the entire wxMaxima code for the calculation of the nodal unknowns based on analytical integration.

The following Module 4.7 shows the entire wxMaxima code for the calculation of the nodal unknowns based on numerical 2×2 integration.

The different results for the displacement at note 3 are shown in Table 4.3.

The Euler-Bernoulli solution for this problem is obtained as $u_{\max} = \frac{-F_0 L^3}{192EI} = -\frac{F_0 a^2}{Eh^3}$. The finite element solution reduces for $\nu \rightarrow 0$ to: $u_{3Z} = -\frac{10}{9}\frac{F_0 a^2}{Eh^3}$.

```
(% i3)  kill(all);
        load("../Library/my_funs.mac")$
        ratprint :false$
        fpprintprec : 6$
(% o0)  done
```

Definition of Classical Kirchhoff Plate Problem: Nodal Coordinates "nodxy", Element Connectivity "elenod", Element Material Properties "elemat", Number of Element Nodes "nen", Thickness Function "thickness", freedom tags list "nodtag", freedom values list "nodval", Numerical or Analytical Calculations Switch "numerical"

```
(% i11) nodxy : [[-a,-a],[a,-a],[a,a],[-a,a]]$

        elenod : [[1,2,3,4]]$

        elemat : [[E, v]]$

        thickness(x,y) := h$
        body_force(x,y) := 0$

        nodtag : [[1,1,1],[0,0,0],[0,0,0],[1,1,1]]$
        nodval : [[0,0,0],[-F_0/2,0,0],[-F_0/2,0,0],[0,0,0]]$

        numerical : false;
        nen : 4$
(numerical)  false
```

Maxima Calculations and Output

```
(% i15) [noddis,Kmod,fmod] :
        KirschhoffPlateElementsSolution(nodxy,elenod,elemat,nodtag,nodval,numerical)$

        coeff : mat[1]*(th^3)/(12*(1-mat[2]^2))$
        noddis : noddis*coeff/D$

        for i : 1 thru length(nodxy) do (
          print(" "),
          print("Displacements at Node ", i),
          print(" ", U_Z, "=", factor(ratsimp(noddis[i][1]))),
          print(" ", φ_X, "=", factor(ratsimp(noddis[i][2]))),
          print(" ", φ_Y, "=", factor(ratsimp(noddis[i][2])))
        )$
```

Displacements at Node 1

$$U_Z = 0$$
$$\varphi_X = 0$$
$$\varphi_Y = 0$$

Displacements at Node 2

$$U_Z = -\frac{2F_0a^2\left(3v^2+2v-6\right)}{3D\,(v-1)\,(2v+3)}$$
$$\varphi_X = \frac{F_0av}{D\,(v-1)\,(2v+3)}$$
$$\varphi_Y = \frac{F_0av}{D\,(v-1)\,(2v+3)}$$

Displacements at Node 3

$$U_Z = -\frac{2F_0a^2\left(3v^2+2v-6\right)}{3D\,(v-1)\,(2v+3)}$$
$$\varphi_X = -\frac{F_0av}{D\,(v-1)\,(2v+3)}$$
$$\varphi_Y = -\frac{F_0av}{D\,(v-1)\,(2v+3)}$$

Displacements at Node 4

$$U_Z = 0$$
$$\varphi_X = 0$$
$$\varphi_Y = 0$$

Module 4.4: Calculation of the nodal unknowns based on analytical integration for the problem shown in Fig. 4.6

Table 4.3 Comparison of the results for the displacements at node 3 depending on the integration approach

Analytical	2 × 2	3 × 3
$u_Z = \frac{5F_0a^2}{2D(2\nu-27)}$	$u_Z = -\frac{F_0a^2}{10D}$	$u_Z = \frac{5F_0a^2}{2D(2\nu-27)}$

The analytical solution for a rectangular plate with central load and all edges built in is given in [5] as:

$$|u_Z|_{\text{analytical}} = \alpha\frac{F_0(4a)^2}{D}, \tag{4.79}$$

where the numerical factor α is given as 0.00560 for a square plate with $\nu = 0.3$. Thus, the analytical solution can be written for the case under consideration as:

(%i3)
```
kill(all);
load("../Library/my_funs.mac")$
ratprint :false$
fpprintprec : 6$
```

(%o0) *done*

Definition of Classical Kirchhoff Plate Problem: Nodal Coordinates "nodxy", Element Connectivity "elenod", Element Material Properties "elemat", Number of Element Nodes "nen", Thickness Function "thickness", freedom tags list "nodtag", freedom values list "nodval", Numerical or Analytical Calculations Switch "numerical"

(%i11)
```
nodxy : [[-a,-a],[a,-a],[a,a],[-a,a]]$

elenod : [[1,2,3,4]]$

elemat : [[E, v]]$

thickness(x,y) := h$
body_force(x,y) := 0$

nodtag : [[1,1,1],[0,0,0],[0,0,0],[1,1,1]]$
nodval : [[0,0,0],[-F_0/2,0,0],[-F_0/2,0,0],[0,0,0]]$

numerical : true;
nqp : 4$
nen : 4$
```

Maxima Calculations and Output

(%i15)
```
[noddis,Kmod,fmod] :
KirschhoffPlateElementsSolution(nodxy,elenod,elemat,nodtag,nodval,numerical)$

coeff : mat[1]*(th^3)/(12*(1-mat[2]^2))$
noddis : noddis*coeff/D$

for i : 1 thru length(nodxy) do (
  print(" "),
  print("Displacements at Node ", i),
  print(" ", U_Z, "=", factor(ratsimp(noddis[i][1]))),
  print(" ", φ_X, "=", factor(ratsimp(noddis[i][2]))),
  print(" ", φ_Y, "=", factor(ratsimp(noddis[i][2])))
)$
```

Displacements at Node 1

$$U_Z = 0$$
$$\varphi_X = 0$$
$$\varphi_Y = 0$$

Displacements at Node 2

$$u_Z = -\frac{2F_0a^2\left(3v^2+2v-6\right)}{3D\,(v-1)\,(2v+3)}$$

$$\varphi_X = \frac{F_0av}{D\,(v-1)\,(2v+3)}$$

$$\varphi_Y = \frac{F_0av}{D\,(v-1)\,(2v+3)}$$

Displacements at Node 3

$$u_Z = -\frac{2F_0a^2\left(3v^2+2v-6\right)}{3D\,(v-1)\,(2v+3)}$$

$$\varphi_X = -\frac{F_0av}{D\,(v-1)\,(2v+3)}$$

$$\varphi_Y = -\frac{F_0av}{D\,(v-1)\,(2v+3)}$$

Displacements at Node 4

$$u_Z = 0$$

$$\varphi_X = 0$$

$$\varphi_Y = 0$$

Module 4.5: Calculation of the nodal unknowns based on numerical integration for the problem shown in Fig. 4.6

$$|u_Z|_{\text{analytical}} = 0.0896 \times \frac{F_0(a)^2}{D}. \tag{4.80}$$

The relative error is obtained as:

$$\frac{||u_Z|_{\text{analytical}} - |u_Z|_{\text{FE}}|}{|u_Z|_{\text{analytical}}} = \frac{0.0947 - 0.0896}{0.0947} = 5.4\%\,. \tag{4.81}$$

```
(% i3)  kill(all);
        load("../Library/my_funs.mac")$
        ratprint :false$
        fpprintprec : 6$

(% o0)  done
```

Definition of Classical Kirchhoff Plate Problem: Nodal Coordinates "nodxy", Element Connectivity "elenod", Element Material Properties "elemat", Number of Element Nodes "nen", Thickness Function "thickness", freedom tags list "nodtag", freedom values list "nodval", Numerical or Analytical Calculations Switch "numerical"

```
(% i13) nodxy : [[0,0],[2*a,0],[2*a,2*a],[0,2*a],[2*a,4*a],[0,4*a],[4*a,0],[4*a,2*a],[4*a,4*a]]$

        elenod : [[1,2,3,4],[4,3,5,6],[2,7,8,3],[3,8,9,5]]$

        elemat : makelist([E, v],9)$

        thickness(x,y) := h$
        body_force(x,y) := 0$

        nodtag : makelist([1,1,1],9)$
        nodtag[3] : [0,0,0]$
        nodval : makelist([0,0,0],9)$
        nodval[3] : [-F_0,0,0]$

        numerical : false$

        nen : 4$
```

Maxima Calculations and Output

```
(% i21) [noddis,Kmod,fmod] : KirschhoffPlateElementsSolution(nodxy,elenod,elemat,
        nodtag,nodval,numerical)$

        coeff : mat[1]*(th^3)/(12*(1-mat[2]^2))$
        noddis : noddis*coeff/D$

        print(" ")$
        print("Deformations at Node ", 3)$
        print(" ", U_Z, "=", factor(ratsimp(noddis[3][1])))$
        print(" ", φ_X, "=", factor(ratsimp(noddis[3][2])))$
        print(" ", φ_Y, "=", factor(ratsimp(noddis[3][2])))$
```

Deformations at Node 3

$$U_Z = \frac{5F_0a^2}{2D(2\nu - 27)}$$

$$\varphi_X = 0$$

$$\varphi_Y = 0$$

Module 4.6: Calculation of the nodal unknowns based on analytical integration for the problem shown in Fig. 4.7

```
(%i3)  kill(all);
       load("../Library/my_funs.mac")$
       ratprint :false$
       fpprintprec : 6$

(%o0)  done
```

Definition of Classical Kirchhoff Plate Problem: Nodal Coordinates "nodxy", Element Connectivity "elenod", Element Material Properties "elemat", Number of Element Nodes "nen", Thickness Function "thickness", freedom tags list "nodtag", freedom values list "nodval", Numerical or Analytical Calculations Switch "numerical"

```
(%i13) nodxy : [[0,0],[2*a,0],[2*a,2*a],[0,2*a],[2*a,4*a],[0,4*a],[4*a,0],[4*a,2*a],[4*a,4*a]]$

       elenod : [[1,2,3,4],[4,3,5,6],[2,7,8,3],[3,8,9,5]]$

       elemat : makelist([E, v],9)$

       thickness(x,y) := h$
       body_force(x,y) := 0$

       nodtag : makelist([1,1,1],9)$
       nodtag[3] : [0,0,0]$
       nodval : makelist([0,0,0],9)$
       nodval[3] : [-F_0,0,0]$

       numerical : true;
       nqp : 4$
       nen : 4$
```

Maxima Calculations and Output

```
(%i21) [noddis,Kmod,fmod] : KirschhoffPlateElementsSolution(nodxy,elenod,elemat,
       nodtag,nodval,numerical)$

       coeff : mat[1]*(th^3)/(12*(1-mat[2]^2))$
       noddis : noddis*coeff/D$

       print(" ")$
       print("Deformations at Node ", 3)$
       print(" ", U_Z, "=", factor(ratsimp(noddis[3][1])))$
       print(" ", φ_X, "=", factor(ratsimp(noddis[3][2])))$
       print(" ", φ_Y, "=", factor(ratsimp(noddis[3][2])))$
```

Deformations at Node 3

$$U_Z = -\frac{F_0 a^2}{10D}$$

$$\varphi_X = 0$$

$$\varphi_Y = 0$$

Module 4.7: Calculation of the nodal unknowns based on numerical 2×2 integration for the problem shown in Fig. 4.7

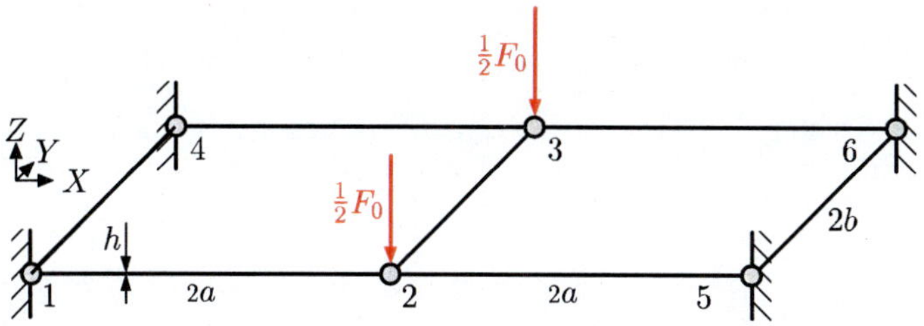

Fig. 4.8 Plate problem with two edges fixed

4.5 Symmetry solution for a plate fixed at all four edges

Reconsider Example 4.4 and derive the solution under consideration of the symmetry of the problem, i.e., based on a single element ($2a \times 2a \times h$) and corresponding boundary conditions. Base your solution on analytical integration.

4.5 Solution
The following Module 4.8 shows the entire wxMaxima code for the calculation of the nodal unknowns based on analytical integration.

4.6 Two-element example of a plate fixed at two edges

Given is a classical plate structure which is fixed at two sides, see Fig. 4.8. The side lengths of the entire structure are equal to $4a \times 2b$. The plate is loaded by two single forces $\frac{1}{2}F_0$ acting in the middle of the plate structure. The material is described based on the engineering constants Young's modulus E and Poisson's ratio ν.

Use two classical plate elements (each $2a \times 2b \times h$) in the following to model the problem and to calculate the nodal unknowns in the middle of the plate structure. Simplify the results for the special case $\nu \to 0$ and compare these results with the Euler-Bernoulli solution.

4.6 Solution
The following Module 4.9 shows the entire wxMaxima code for the calculation of the nodal unknowns based on analytical integration. It should be noted here that a 2×2 numerical integration scheme gives the same results as in the case of analytical integration.

The special case $\nu \to 0$ gives $u_Z = -\frac{2a^3 F_0}{Ebh^3}$ and $\varphi_X = \varphi_Y = 0$ which is equal to the Euler-Bernoulli solution.

```
(% i3)  kill(all);
        load("../Library/my_funs.mac")$
        ratprint :false$
        fpprintprec : 6$

(% o0)  done
```

Definition of Classical Kirchhoff Plate Problem: Nodal Coordinates "nodxy", Element Connectivity "elenod", Element Material Properties "elemat", Number of Element Nodes "nen", Thickness Function "thickness", freedom tags list "nodtag", freedom values list "nodval", Numerical or Analytical Calculations Switch "numerical"

```
(% i13) nodxy : [[0,0],[2*a,0],[2*a,2*a],[0,2*a]]$

        elenod : [[1,2,3,4]]$

        elemat : makelist([E, v],length(elenod))$

        thickness(x,y) := h$
        body_force(x,y) := 0$

        nodtag : makelist([1,1,1],length(nodxy))$
        nodtag[3] : [0,1,1]$
        nodval : makelist([0,0,0],length(nodxy))$
        nodval[3] : [-F_0/4,0,0]$

        numerical : false$
        nen : 4$
```

Maxima Calculations and Output

```
(% i21) [noddis,Kmod,fmod] : KirschhoffPlateElementsSolution(nodxy,elenod,elemat,
        nodtag,nodval,numerical)$

        coeff : mat[1]*(th^3)/(12*(1-mat[2]^2))$
        noddis : noddis*coeff/D$

        print(" ")$
        print("Displacements and rotations at Node ", 3)$
        print(" ", U_Z, "=", factor(ratsimp(noddis[3][1])))$
        print(" ", φ_X, "=", factor(ratsimp(noddis[3][2])))$
        print(" ", φ_Y, "=", factor(ratsimp(noddis[3][2])))$
```

Displacements and rotations at Node at Node 3

$$U_Z = \frac{5F_0a^2}{2D(2\nu-27)}$$

$$\varphi_X = 0$$

$$\varphi_Y = 0$$

Module 4.8: Calculation of the nodal unknowns based on analytical integration for the problem shown in Fig. 4.7 under consideration of the symmetry

```
(% i3)  kill(all);
        load("../Library/my_funs.mac")$
        ratprint :false$
        fpprintprec : 6$
(% o0)  done
```

Definition of Classical Kirchhoff Plate Problem: Nodal Coordinates "nodxy", Element Connectivity "elenod", Element Material Properties "elemat", Number of Element Nodes "nen", Thickness Function "thickness", freedom tags list "nodtag", freedom values list "nodval", Numerical or Analytical Calculations Switch "numerical"

```
(% i15) nodxy : [[0,0],[2*a,0],[2*a,2*b],[0,2*b],[4*a,0],[4*a,2*b]]$
        elenod : [[1,2,3,4],[2,5,6,3]]$
        elemat : makelist([E, v],length(elenod))$

        thickness(x,y) := h$
        body_force(x,y) := 0$

        nodtag : makelist([1,1,1],length(nodxy))$
        nodtag[2] : [0,0,0]$
        nodtag[3] : [0,0,0]$
        nodval : makelist([0,0,0],length(nodxy))$
        nodval[2] : [-F_0/2,0,0]$
        nodval[3] : [-F_0/2,0,0]$

        numerical : false$
        nen : 4$
```

Maxima Calculations and Output

```
(% i20) [noddis,Kmod,fmod] : KirschhoffPlateElementsSolution(nodxy,elenod,elemat,
        nodtag,nodval,numerical)$

        coeff : mat[1]*(th^3)/(12*(1-mat[2]^2))$
        noddis : noddis*coeff/D$

        postlist : [2,3]$
        for i in postlist do (
          print(" "),
          print("Displacements and rotations at Node ", i),
          print(" ", U_Z, "=", factor(ratsimp(noddis[i][1]))),
          print(" ", φ_X, "=", factor(ratsimp(noddis[i][2]))),
          print(" ", φ_Y, "=", factor(ratsimp(noddis[i][3])))
        )$
```

Displacements and rotations at Node at Node 2

$$U_Z = -\frac{F_0 a^3 \left(b^2 v - b^2 - 2a^2\right)}{3Db \left(a^2 v^2 + 2b^2 v - 2b^2 - 4a^2\right)}$$

$$\varphi_X = -\frac{F_0 a^3 v}{2D \left(a^2 v^2 + 2b^2 v - 2b^2 - 4a^2\right)}$$

$$\varphi_Y = 0$$

Displacements and rotations at Node at Node 3

$$u_Z = -\frac{F_0 a^3 \left(b^2 \nu - b^2 - 2a^2\right)}{3Db \left(a^2 \nu^2 + 2b^2 \nu - 2b^2 - 4a^2\right)}$$

$$\varphi_X = \frac{F_0 a^3 \nu}{2D \left(a^2 \nu^2 + 2b^2 \nu - 2b^2 - 4a^2\right)}$$

$$\varphi_Y = 0$$

Module 4.9: Calculation of the nodal unknowns based on analytical integration for the problem shown in Fig. 4.8

4.7 Four-element example of a plate fixed at all four edges with constant distributed load

Given is a classical plate which is fixed at all four sides, see Fig. 4.9. The side lengths are equal to $4a$. The plate is loaded by a constant distributed load q_0. The material is described based on the engineering constants Young's modulus E and Poisson's ratio ν.

Use four classical plate elements (each $2a \times 2a \times h$) in the following to model the problem and to calculate the nodal unknowns in the middle of the plate.

Use the analytical solution provided in [5] to benchmark the finite element solution.

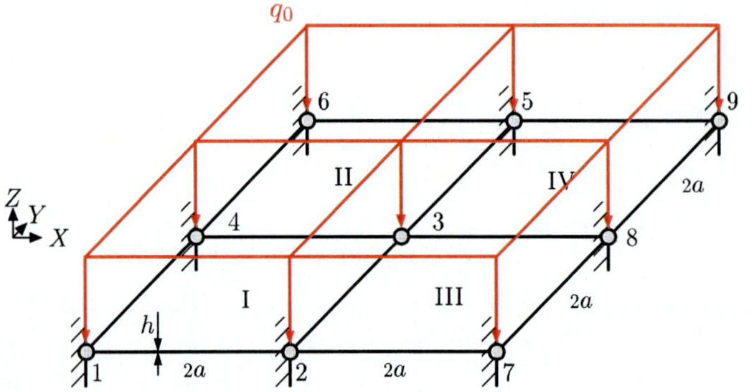

Fig. 4.9 Plate problem with four edges fixed and constant distributed load

```
(%i3)  kill(all);
       load("../Library/my_funs.mac")$
       ratprint :false$
       fpprintprec : 6$

(%o0)  done
```

Definition of Classical Kirchhoff Plate Problem: Nodal Coordinates "nodxy", Element Connectivity "elenod", Element Material Properties "elemat", Number of Element Nodes "nen", Thickness Function "thickness", freedom tags list "nodtag", freedom values list "nodval", Numerical or Analytical Calculations Switch "numerical"

```
(%i13) nodxy : [[0,0],[2*a,0],[2*a,2*a],[0,2*a],[2*a,4*a],[0,4*a],[4*a,0],[4*a,2*a],[4*a,4*a]]$

       elenod : [[1,2,3,4],[4,3,5,6],[2,7,8,3],[3,8,9,5]]$

       elemat : makelist([E, v],length(elenod))$

       thickness(x,y) := h$
       body_force(x,y) := q_0$

       nodtag : makelist([1,1,1],length(nodxy))$
       nodtag[3] : [0,0,0]$
       nodval : makelist([0,0,0],length(nodxy))$

       numerical : false$
       nen : 4$
```

Maxima Calculations and Output

```
(%i14) [noddis,Kmod,fmod] : KirschhoffPlateElementsSolution(nodxy,elenod,elemat,
       nodtag,nodval,numerical)$

       coeff : mat[1]*(th^3)/(12*(1-mat[2]^2))$
       noddis : noddis*coeff/D$

       postlist : [3]$
       for i in postlist do (
         print(" "),
         print("Displacements and rotations at Node ", i),
         print(" ", U_Z, "=", factor(ratsimp(noddis[i][1]))),
         print(" ", φ_X, "=", factor(ratsimp(noddis[i][2]))),
         print(" ", φ_Y, "=", factor(ratsimp(noddis[i][3])))
       )$
```

Displacements and rotations at Node at Node 3

$$U_Z = -\frac{10 q_0 a^4}{D(2v - 27)}$$

$$\varphi_X = 0$$

$$\varphi_Y = 0$$

Module 4.10: Calculation of the nodal unknowns based on analytical integration for the problem shown in Fig. 4.9

4.7 Solution

The following Module 4.10 shows the entire wxMaxima code for the calculation of the nodal unknowns based on analytical integration.

The analytical solution for a rectangular plate with central load and all edges built in is given in [5] as:

$$|u_Z|_{\text{analytical}} = \alpha \frac{q_0(4a)^4}{D}, \tag{4.82}$$

where the numerical factor α is given as 0.00126 for a square plate with $\nu = 0.3$. Thus, the analytical solution can be written for the case under consideration as:

$$|u_Z|_{\text{analytical}} = 0.3226 \times \frac{q_0 a^4}{D}. \tag{4.83}$$

The relative error is obtained as:

$$\frac{||u_Z|_{\text{analytical}} - |u_Z|_{\text{FE}}|}{|u_Z|_{\text{analytical}}} = \left|\frac{0.3226 - 0.3788}{0.3226}\right| = 17.42\%\,. \tag{4.84}$$

References

1. Öchsner A (2014) Elasto-plasticity of frame structure elements: modeling and simulation of rods and beams. Springer, Berlin
2. Öchsner A, Makvandi M (2019) Finite elements using maxima: theory and routines for rods and beams. Springer, Cham
3. Öchsner A (2020) Computational statics and dynamics - an introduction based on the finite element method. Springer, Singapore
4. Reddy JN (2006) An introduction to the finite element method. McGraw Hill, Singapore
5. Timoshenko S, Woinowsky-Krieger S (1959) Theory of plates and shells. McGraw-Hill Book Company, New York

Chapter 5
Shear Deformable Plate Elements

5.1 Calculation of the Elemental Stiffness Matrix

A thick plate is similarly defined as a thin plate (see Fig. 4.1). However, the condition that the thickness h is *much* smaller than the planar dimensions (a and b) is weakened. The thickness is still smaller than a and b and not in the same range, see Fig. 5.1. The case of $h \approx a \approx b$ would rather refer to a three-dimensional element (see [2]). The thick plate can be seen as a two-dimensional extension or generalization of the Timoshenko beam and is also called the Reissner-Mindlin plate[1] in the finite element context.

Let us note here that the constitutive equation for a thick plate can be expressed in matrix notation as [2]

$$\begin{bmatrix} M_x^{\mathrm{n}} \\ M_y^{\mathrm{n}} \\ M_{xy}^{\mathrm{n}} \\ Q_x^{\mathrm{n}} \\ Q_y^{\mathrm{n}} \end{bmatrix} = \left[\begin{array}{c|c} \underbrace{\frac{Eh^3}{12(1-\nu^2)}}_{D_\mathrm{b}} \begin{bmatrix} 1 & \nu & 0 \\ \nu & 1 & 0 \\ 0 & 0 & \frac{1-\nu}{2} \end{bmatrix} & \begin{bmatrix} 0 & 0 \\ 0 & 0 \\ 0 & 0 \end{bmatrix} \\ \hline \begin{bmatrix} 0 & 0 & 0 \\ 0 & 0 & 0 \end{bmatrix} & \underbrace{k_\mathrm{s} G h}_{D_\mathrm{s}} \begin{bmatrix} 1 & 0 \\ 0 & 1 \end{bmatrix} \end{array}\right] \begin{bmatrix} \frac{\partial \phi_y}{\partial x} \\ -\frac{\partial \phi_x}{\partial y} \\ \frac{\partial \phi_y}{\partial y} - \frac{\partial \phi_x}{\partial x} \\ \phi_y + \frac{\partial u_z}{\partial x} \\ -\phi_x + \frac{\partial u_z}{\partial y} \end{bmatrix}, \tag{5.1}$$

or symbolically as

$$\boldsymbol{s}^* = \boldsymbol{D}^* \boldsymbol{e}, \tag{5.2}$$

where $\boldsymbol{D}^*$ is the plate elasticity matrix,[2] $\boldsymbol{s}^*$ are the generalized stresses, $\boldsymbol{e}$ are the generalized strains. Furthermore, E is Young's modulus, ν is Poisson's ratio, G is the shear modulus, k_s is the shear correction factor, M^n is a length-specific moment, and Q^n is a length-specific shear force.

---

[1] Strictly speaking, there is a small difference between the plate theory according to Reissner [3] and Mindlin [1] and only for zero Poisson's ratio both derivations are the same.

[2] This plate elasticity matrix should not be confused with the compliance matrix which is represented by the same symbol.

A. Öchsner and R. Makvandi, *Plane Finite Elements for Two-Dimensional Problems*,
https://doi.org/10.1007/978-3-030-89550-1_5

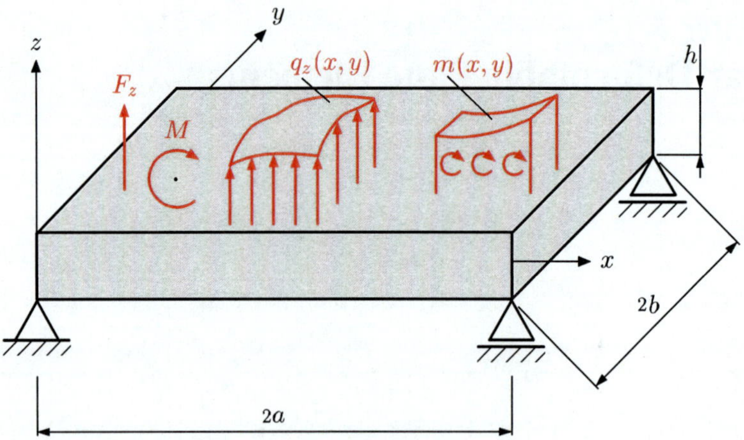

Fig. 5.1 General configuration for a thick plate problem

5.1.1 Elements with Four Nodes

Let us consider in the following the four-node thick plate element as shown in Fig. 5.2. Important for the following derivations is that the node numbering is counterclockwise.

The general expression for the elemental stiffness matrix $\boldsymbol{K}^{\mathrm{e}}$ of a four-node thick plate element with the plate elasticity matrix $\boldsymbol{D}^*$ can be expressed as

$$\boldsymbol{K}^{\mathrm{e}} = \int_A \underbrace{\left(\boldsymbol{\mathcal{L}}_1 \boldsymbol{N}^{\mathrm{T}}\right)^{\mathrm{T}}}_{\boldsymbol{B}} \boldsymbol{D}^* \underbrace{\left(\boldsymbol{\mathcal{L}}_1 \boldsymbol{N}^{\mathrm{T}}\right)}_{\boldsymbol{B}^{\mathrm{T}}} \mathrm{d}A\,, \tag{5.3}$$

where the $\boldsymbol{B}$-matrix contains the first-order derivatives of the interpolation functions N_i $(i = 1, \ldots, 4)$.[3] Its transposed can be expressed as:

$$\boldsymbol{B}^{\mathrm{T}} = \boldsymbol{\mathcal{L}}_1 \boldsymbol{N}^{\mathrm{T}} = \begin{bmatrix} 0 & 0 & \frac{\partial}{\partial x} \\ 0 & -\frac{\partial}{\partial y} & 0 \\ 0 & -\frac{\partial}{\partial x} & \frac{\partial}{\partial y} \\ \frac{\partial}{\partial x} & 0 & 1 \\ \frac{\partial}{\partial y} & -1 & 0 \end{bmatrix} \left[\begin{array}{ccc|c|ccc} N_1 & 0 & 0 & \cdots & N_4 & 0 & 0 \\ 0 & N_1 & 0 & \cdots & 0 & N_4 & 0 \\ 0 & 0 & N_1 & \cdots & 0 & 0 & N_4 \end{array}\right] = \tag{5.4}$$

[3] Each node has three degrees of freedom and we assume here that $N_{\phi_y} = N_{\phi_x} = N_u$.

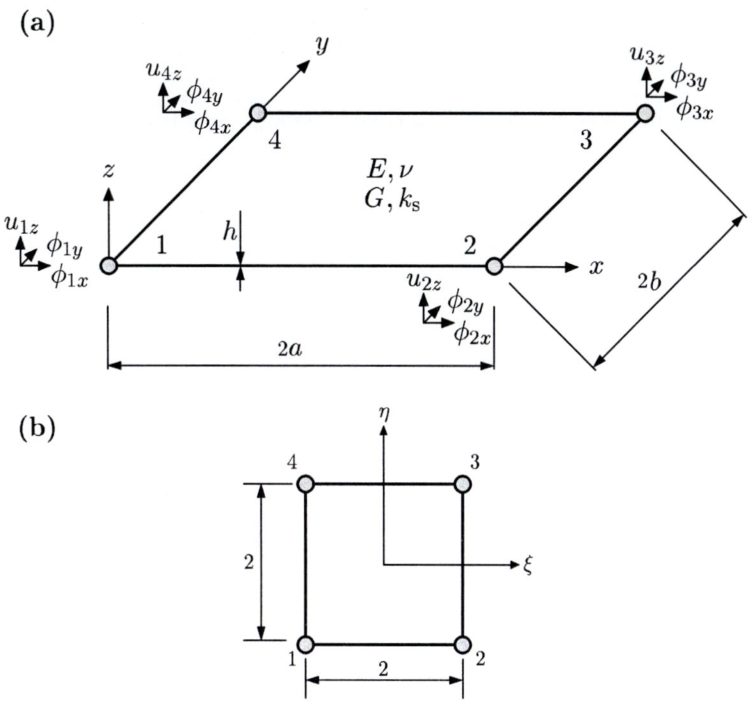

Fig. 5.2 Rectangular four-node thick plate element: **a** Cartesian and **b** parametric space

$$
= \left[\begin{array}{ccc|c|ccc}
0 & 0 & \frac{\partial N_1}{\partial x} & & 0 & 0 & \frac{\partial N_4}{\partial x} \\
0 & -\frac{\partial N_1}{\partial y} & 0 & \cdots & 0 & -\frac{\partial N_4}{\partial y} & 0 \\
0 & -\frac{\partial N_1}{\partial x} & \frac{\partial N_1}{\partial y} & \cdots & 0 & -\frac{\partial N_4}{\partial x} & \frac{\partial N_4}{\partial y} \\
\frac{\partial N_1}{\partial x} & 0 & N_1 & \cdots & \frac{\partial N_4}{\partial x} & 0 & N_n \\
\frac{\partial N_1}{\partial y} & -N_1 & 0 & & \frac{\partial N_4}{\partial y} & -N_4 & 0
\end{array}\right] = \boldsymbol{B}^{\mathrm{T}} . \tag{5.5}
$$

The integration for the element stiffness matrix given in Eq. (5.3) is approximated by numerical integration. To this end, the coordinates (x, y) are transformed to the natural coordinates (unit space) (ξ, η) where each coordinate ranges from -1 to 1. In the scope of the coordinate transformation, attention must be paid to the derivatives. For example, the derivative of the interpolation functions with respect to the x-coordinate is transformed in the following way:

$$\frac{\partial N_i}{\partial x} \rightarrow \frac{\partial N_i}{\partial \xi}\frac{\partial \xi}{\partial x} + \frac{\partial N_i}{\partial \eta}\frac{\partial \eta}{\partial x}. \tag{5.6}$$

Furthermore, the coordinate transformation requires that $\mathrm{d}A = \mathrm{d}x\mathrm{d}y \rightarrow \mathrm{d}A' = J\mathrm{d}\xi\mathrm{d}\eta$, where J is the Jacobian:

$$J = \left|\frac{\partial(x, y)}{\partial(\xi, \eta)}\right| = \frac{\partial x}{\partial \xi} \times \frac{\partial y}{\partial \eta} - \frac{\partial x}{\partial \eta} \times \frac{\partial y}{\partial \xi}. \tag{5.7}$$

Assuming bilinear interpolation functions, the relations from Chap. 3 can be taken, see Eqs. (3.12)–(3.15). Thus, the interpolations functions read as

$$N_1(\xi, \eta) = \frac{1}{4}(1 - \xi - \eta + \xi\eta) = \frac{1}{4}(1 - \xi)(1 - \eta), \tag{5.8}$$

$$N_2(\xi, \eta) = \frac{1}{4}(1 + \xi - \eta - \xi\eta) = \frac{1}{4}(1 + \xi)(1 - \eta), \tag{5.9}$$

$$N_3(\xi, \eta) = \frac{1}{4}(1 + \xi + \eta + \xi\eta) = \frac{1}{4}(1 + \xi)(1 + \eta), \tag{5.10}$$

$$N_4(\xi, \eta) = \frac{1}{4}(1 - \xi + \eta - \xi\eta) = \frac{1}{4}(1 - \xi)(1 + \eta). \tag{5.11}$$

The derivatives with respect to the parametric coordinates can easily be obtained as:

$$\frac{\partial N_1(\xi, \eta)}{\partial \xi} = \frac{1}{4}(-1 + \eta) \quad ; \quad \frac{\partial N_1(\xi, \eta)}{\partial \eta} = \frac{1}{4}(-1 + \xi), \tag{5.12}$$

$$\frac{\partial N_2(\xi, \eta)}{\partial \xi} = \frac{1}{4}(+1 - \eta) \quad ; \quad \frac{\partial N_2(\xi, \eta)}{\partial \eta} = \frac{1}{4}(-1 - \xi), \tag{5.13}$$

$$\frac{\partial N_3(\xi, \eta)}{\partial \xi} = \frac{1}{4}(+1 + \eta) \quad ; \quad \frac{\partial N_3(\xi, \eta)}{\partial \eta} = \frac{1}{4}(+1 + \xi), \tag{5.14}$$

$$\frac{\partial N_4(\xi, \eta)}{\partial \xi} = \frac{1}{4}(-1 - \eta) \quad ; \quad \frac{\partial N_4(\xi, \eta)}{\partial \eta} = \frac{1}{4}(+1 - \xi). \tag{5.15}$$

Geometrical Derivatives

Let us assume the same interpolation for the global x- and y-coordinate as for the displacement (*isoparametric* element formulation), i.e. $\overline{N}_i = N_i$:

$$x(\xi,\eta) = \overline{N}_1(\xi,\eta) \times x_1 + \overline{N}_2(\xi,\eta) \times x_2 + \overline{N}_3(\xi,\eta) \times x_3 + \overline{N}_4(\xi,\eta) \times x_4\,, \tag{5.16}$$

$$y(\xi,\eta) = \overline{N}_1(\xi,\eta) \times y_1 + \overline{N}_2(\xi,\eta) \times y_2 + \overline{N}_3(\xi,\eta) \times y_3 + \overline{N}_4(\xi,\eta) \times y_4\,. \tag{5.17}$$

Remark: the *global* coordinates of the nodes $1, \ldots, 4$ can be used for $x_1, \ldots, x_4$ and $y_1, \ldots, y_4$.

Thus, the geometrical derivatives can easily be obtained as:

$$\frac{\partial x}{\partial \xi} = \frac{1}{4}\Big((-1+\eta)x_1 + (1-\eta)x_2 + (1+\eta)x_3 + (-1-\eta)x_4\Big)\,, \tag{5.18}$$

$$\frac{\partial y}{\partial \xi} = \frac{1}{4}\Big((-1+\eta)y_1 + (1-\eta)y_2 + (1+\eta)y_3 + (-1-\eta)y_4\Big)\,, \tag{5.19}$$

$$\frac{\partial x}{\partial \eta} = \frac{1}{4}\Big((-1+\xi)x_1 + (-1-\xi)x_2 + (1+\xi)x_3 + (1-\xi)x_4\Big)\,, \tag{5.20}$$

$$\frac{\partial y}{\partial \eta} = \frac{1}{4}\Big((-1+\xi)y_1 + (-1-\xi)y_2 + (1+\xi)y_3 + (1-\xi)y_4\Big)\,. \tag{5.21}$$

The calculation of the derivatives of the interpolation functions (see Eq. (5.6)) requires, however, the geometrical derivatives of the natural coordinates (ξ, η) with respect to the physical coordinates (x, y). These relations can be easily obtained from Eqs. (5.18)–(5.21) under consideration of the following relationships:

$$\frac{\partial \xi}{\partial x} = +\frac{1}{\frac{\partial x}{\partial \xi}\frac{\partial y}{\partial \eta} - \frac{\partial x}{\partial \eta}\frac{\partial y}{\partial \xi}} \times \frac{\partial y}{\partial \eta}\,, \tag{5.22}$$

$$\frac{\partial \xi}{\partial y} = -\frac{1}{\frac{\partial x}{\partial \xi}\frac{\partial y}{\partial \eta} - \frac{\partial x}{\partial \eta}\frac{\partial y}{\partial \xi}} \times \frac{\partial x}{\partial \eta}\,, \tag{5.23}$$

$$\frac{\partial \eta}{\partial x} = -\frac{1}{\frac{\partial x}{\partial \xi}\frac{\partial y}{\partial \eta} - \frac{\partial x}{\partial \eta}\frac{\partial y}{\partial \xi}} \times \frac{\partial y}{\partial \xi}\,, \tag{5.24}$$

$$\frac{\partial \eta}{\partial y} = +\frac{1}{\frac{\partial x}{\partial \xi}\frac{\partial y}{\partial \eta} - \frac{\partial x}{\partial \eta}\frac{\partial y}{\partial \xi}} \times \frac{\partial x}{\partial \xi}\,. \tag{5.25}$$

Based on the derived equations, the triple matrix product $\boldsymbol{B}\boldsymbol{C}^*\boldsymbol{B}^{\mathrm{T}}$ (see Eq. (5.3)) can be numerically calculated to obtain the stiffness matrix.

Let us summarize here the major steps which are required to calculate the elemental stiffness matrix.

❶ Introduce an elemental coordinate system (x, y).

❷ Express the coordinates (x_i, y_i) of the corner nodes i $(i = 1, \ldots, 4)$ in this elemental coordinate system.

❸ Calculate the partial derivatives of the Cartesian (x, y) coordinates with respect to the natural (ξ, η) coordinates, see Eqs. (5.18)–(5.21):

$$\frac{\partial x}{\partial \xi} = x_\xi = \frac{1}{4}\Big((-1+\eta)x_1 + (1-\eta)x_2 + (1+\eta)x_3 + (-1-\eta)x_4\Big),$$

$$\vdots$$

$$\frac{\partial y}{\partial \eta} = y_\eta = \frac{1}{4}\Big((-1+\xi)y_1 + (-1-\xi)y_2 + (1+\xi)y_3 + (1-\xi)y_4\Big).$$

❹ Calculate the first-order partial derivatives of the natural (ξ, η) coordinates with respect to the Cartesian (x, y) coordinates according to Eqs. (5.22)–(5.25):

$$\frac{\partial \xi}{\partial x} = +\frac{1}{\frac{\partial x}{\partial \xi}\frac{\partial y}{\partial \eta} - \frac{\partial x}{\partial \eta}\frac{\partial y}{\partial \xi}} \times \frac{\partial y}{\partial \eta}, \quad \frac{\partial \xi}{\partial y} = -\frac{1}{\frac{\partial x}{\partial \xi}\frac{\partial y}{\partial \eta} - \frac{\partial x}{\partial \eta}\frac{\partial y}{\partial \xi}} \times \frac{\partial x}{\partial \eta},$$

$$\frac{\partial \eta}{\partial x} = -\frac{1}{\frac{\partial x}{\partial \xi}\frac{\partial y}{\partial \eta} - \frac{\partial x}{\partial \eta}\frac{\partial y}{\partial \xi}} \times \frac{\partial y}{\partial \xi}, \quad \frac{\partial \eta}{\partial y} = +\frac{1}{\frac{\partial x}{\partial \xi}\frac{\partial y}{\partial \eta} - \frac{\partial x}{\partial \eta}\frac{\partial y}{\partial \xi}} \times \frac{\partial x}{\partial \xi}.$$

❺ Calculate the $\boldsymbol{B}$-matrix and its transposed, see Eq. (5.5):

$$\boldsymbol{B}^{\mathrm{T}} = \left[\begin{array}{ccc|c|ccc} 0 & 0 & \frac{\partial N_1}{\partial x} & & 0 & 0 & \frac{\partial N_4}{\partial x} \\ 0 & -\frac{\partial N_1}{\partial y} & 0 & & 0 & -\frac{\partial N_4}{\partial y} & 0 \\ 0 & -\frac{\partial N_1}{\partial x} & \frac{\partial N_1}{\partial y} & \cdots & 0 & -\frac{\partial N_4}{\partial x} & \frac{\partial N_4}{\partial y} \\ \frac{\partial N_1}{\partial x} & 0 & N_1 & & \frac{\partial N_4}{\partial x} & 0 & N_4 \\ \frac{\partial N_1}{\partial y} & -N_1 & 0 & & \frac{\partial N_4}{\partial y} & -N_4 & 0 \end{array}\right],$$

where the first-order partial derivatives are $\frac{\partial N_1(\xi,\eta)}{\partial x} = \frac{\partial N_1}{\partial \xi}\frac{\partial \xi}{\partial x} + \frac{\partial N_1}{\partial \eta}\frac{\partial \eta}{\partial x}, \ldots$ and the derivatives of the interpolation functions are given in Eqs. (5.8)–(5.11), i.e., $\frac{\partial N_1}{\partial \xi} = \frac{1}{4}(-1+\eta), \ldots$

❻ Calculate the triple matrix product $\boldsymbol{B}\boldsymbol{D}^*\boldsymbol{B}^{\mathrm{T}}$, where the plate elasticity matrix $\boldsymbol{D}^*$ is given in Eq. (5.1).

❼ Perform the numerical integration based on n integration points:

$$\int_A (\boldsymbol{B}\boldsymbol{D}^*\boldsymbol{B}^{\mathrm{T}})\mathrm{d}A = \sum_{i=1}^{n} \boldsymbol{B}\boldsymbol{D}^*\boldsymbol{B}^{\mathrm{T}} J \times w_i\Big|_{\xi_i,\eta_i}.$$

❽ $\boldsymbol{K}^{\mathrm{e}}$ obtained.

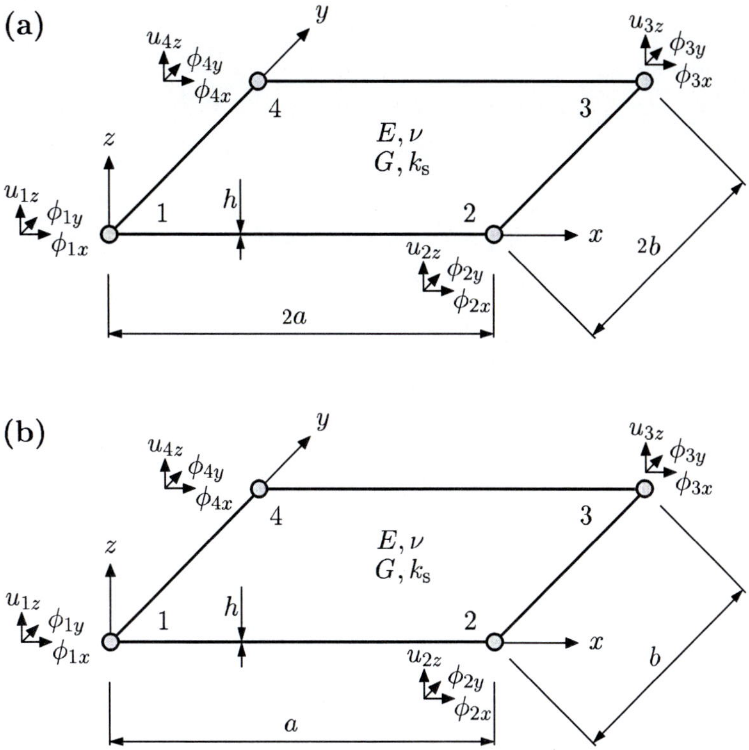

Fig. 5.3 Rectangular four-node thick plate elements with different dimensions: **a** $2a \times 2b \times h$ and **b** $a \times b \times h$

5.1.2 Maxima Modules for Stiffness Matrices

5.1.3 Maxima Examples to Calculate the Stiffness Matrix

5.1 Example: Stiffness matrices for rectangular four-node thick plate elements
Given are different rectangular four-node thick plate elements with dimensions $2a \times 2b \times h$ (see Fig. 5.3a) and $a \times b \times h$ (see Fig. 5.3b).

Derive the expressions for the elemental stiffness matrices based on (a) analytical integration and (b) numerical one-point, 2×2, and 3×3 integration rules.

5.1 Solution
The following Module 5.1 shows the entire wxMaxima code for the calculation of the elemental stiffness matrix (geometry: $2a \times 2b \times h$) based on analytical integration.

```
(%i3)  kill(all);
       load("../Library/my_funs.mac")$
       ratprint :false$
       fpprintprec : 6$

(%o0)  done
```

Definition of Timoshenko Plate Stiffness: Nodal Coordinates "ncoor", Material Properties "mat", Thickness Function "thickness", Number of Element Nodes "nen"

```
(%i7)  ncoor : [[0,0],[2*a,0],[2*a,2*b],[0,2*b]]$

       mat : [E, v, G, Ks]$

       thickness(x,y) := h$

       nen : 4$
```

Maxima Calculations and Output

```
(%i12) Ke :
       k_el_2D_Reissner_Mindlin_plate_analytical(ncoor, mat, nen)$

       print(" ")$
       print("Element stiffness matrix:")$
       print(" ")$
       print(factor(ratsimp(Ke)))$
```

Element stiffness matrix:

$$\begin{bmatrix} \frac{D_s\left(b^2+a^2\right)}{3ab} & \frac{D_s a}{3} & \cdots \\ \frac{D_s a}{3} & -\frac{3D_b b^2 v-8D_s a^2 b^2-3D_b b^2-6D_b a^2}{18ab} & \cdots \\ -\frac{D_s b}{3} & -\frac{D_b (v+1)}{8} & \cdots \\ -\frac{D_s\left(2b^2-a^2\right)}{6ab} & \frac{D_s a}{6} & \cdots \\ \frac{D_s a}{6} & \frac{3D_b b^2 v+4D_s a^2 b^2-3D_b b^2+3D_b a^2}{18ab} & \cdots \\ -\frac{D_s b}{3} & \frac{D_b (3v-1)}{8} & \cdots \\ -\frac{D_s\left(b^2+a^2\right)}{6ab} & -\frac{D_s a}{6} & \cdots \\ \vdots & \vdots & \vdots \end{bmatrix}$$

Module 5.1: Calculation of the elemental stiffness matrix based on analytical integration for the problem shown in Fig. 5.3a (geometry: $2a \times 2b \times h$). The complete stiffness matrix is shown in Eq. (5.26)

$$
\boldsymbol{K}^{\mathrm{e}} =
\begin{bmatrix}
\frac{D_{\mathrm{s}}\left(b^2+a^2\right)}{3ab} & \frac{D_{\mathrm{s}}a}{3} & -\frac{D_{\mathrm{s}}b}{3} & -\frac{D_{\mathrm{s}}\left(2b^2-a^2\right)}{6ab} & \cdots \\
\frac{D_{\mathrm{s}}a}{3} & -\frac{3D_{\mathrm{b}}b^2v-8D_{\mathrm{s}}a^2b^2-3D_{\mathrm{b}}b^2-6D_{\mathrm{b}}a^2}{18ab} & -\frac{D_{\mathrm{b}}(v+1)}{8} & \frac{D_{\mathrm{s}}a}{6} & \cdots \\
-\frac{D_{\mathrm{s}}b}{3} & -\frac{D_{\mathrm{b}}(v+1)}{8} & -\frac{3D_{\mathrm{b}}a^2v-8D_{\mathrm{s}}a^2b^2-6D_{\mathrm{b}}b^2-3D_{\mathrm{b}}a^2}{18ab} & \frac{D_{\mathrm{s}}b}{3} & \cdots \\
-\frac{D_{\mathrm{s}}\left(2b^2-a^2\right)}{6ab} & \frac{D_{\mathrm{s}}a}{6} & \frac{D_{\mathrm{s}}b}{3} & \frac{D_{\mathrm{s}}\left(b^2+a^2\right)}{3ab} & \cdots \\
\frac{D_{\mathrm{s}}a}{6} & \frac{3D_{\mathrm{b}}b^2v+4D_{\mathrm{s}}a^2b^2-3D_{\mathrm{b}}b^2+3D_{\mathrm{b}}a^2}{18ab} & -\frac{D_{\mathrm{b}}(3v-1)}{8} & \frac{D_{\mathrm{s}}a}{3} & \cdots \\
-\frac{D_{\mathrm{s}}b}{3} & \frac{D_{\mathrm{b}}(3v-1)}{8} & -\frac{3D_{\mathrm{b}}a^2v-8D_{\mathrm{s}}a^2b^2+12D_{\mathrm{b}}b^2-3D_{\mathrm{b}}a^2}{36ab} & \frac{D_{\mathrm{s}}b}{3} & \cdots \\
-\frac{D_{\mathrm{s}}\left(b^2+a^2\right)}{6ab} & -\frac{D_{\mathrm{s}}a}{6} & \frac{D_{\mathrm{s}}b}{6} & \frac{D_{\mathrm{s}}\left(b^2-2a^2\right)}{6ab} & \cdots \\
\frac{D_{\mathrm{s}}a}{6} & \frac{3D_{\mathrm{b}}b^2v+4D_{\mathrm{s}}a^2b^2-3D_{\mathrm{b}}b^2-6D_{\mathrm{b}}a^2}{36ab} & \frac{D_{\mathrm{b}}(v+1)}{8} & \frac{D_{\mathrm{s}}a}{3} & \cdots \\
-\frac{D_{\mathrm{s}}b}{6} & \frac{D_{\mathrm{b}}(v+1)}{8} & \frac{3D_{\mathrm{b}}a^2v+4D_{\mathrm{s}}a^2b^2-6D_{\mathrm{b}}b^2-3D_{\mathrm{b}}a^2}{36ab} & \frac{D_{\mathrm{s}}b}{6} & \cdots \\
\frac{D_{\mathrm{s}}\left(b^2-2a^2\right)}{6ab} & -\frac{D_{\mathrm{s}}a}{3} & -\frac{D_{\mathrm{s}}b}{6} & -\frac{D_{\mathrm{s}}\left(b^2+a^2\right)}{6ab} & \cdots \\
\frac{D_{\mathrm{s}}a}{3} & -\frac{3D_{\mathrm{b}}b^2v-8D_{\mathrm{s}}a^2b^2-3D_{\mathrm{b}}b^2+12D_{\mathrm{b}}a^2}{36ab} & \frac{D_{\mathrm{b}}(3v-1)}{8} & \frac{D_{\mathrm{s}}a}{6} & \cdots \\
-\frac{D_{\mathrm{s}}b}{6} & -\frac{D_{\mathrm{b}}(3v-1)}{8} & \frac{3D_{\mathrm{b}}a^2v+4D_{\mathrm{s}}a^2b^2+3D_{\mathrm{b}}b^2-3D_{\mathrm{b}}a^2}{18ab} & \frac{D_{\mathrm{s}}b}{6} & \cdots
\end{bmatrix}
$$

$$
\begin{matrix}
\cdots & \frac{D_s a}{6} & -\frac{D_s b}{3} & -\frac{D_s\left(b^2+a^2\right)}{6ab} & \frac{D_s a}{6} & \cdots \\
\cdots & \frac{3D_b b^2 v+4D_s a^2 b^2-3D_b b^2+3D_b a^2}{18ab} & \frac{D_b(3v-1)}{8} & -\frac{D_s a}{6} & \frac{3D_b b^2 v+4D_s a^2 b^2-3D_b b^2-6D_b a^2}{36ab} & \cdots \\
\cdots & -\frac{D_b(3v-1)}{8} & -\frac{3D_b a^2 v-8D_s a^2 b^2+12D_b b^2-3D_b a^2}{36ab} & \frac{D_s b}{6} & \frac{D_b(v+1)}{8} & \cdots \\
\cdots & \frac{D_s a}{3} & \frac{D_s b}{3} & \frac{D_s\left(b^2-2a^2\right)}{6ab} & \frac{D_s a}{3} & \cdots \\
\cdots & -\frac{3D_b b^2 v-8D_s a^2 b^2-3D_b b^2-6D_b a^2}{18ab} & \frac{D_b(v+1)}{8} & -\frac{D_s a}{3} & -\frac{3D_b b^2 v-8D_s a^2 b^2-3D_b b^2+12D_b a^2}{36ab} & \cdots \\
\cdots & \frac{D_b(v+1)}{8} & -\frac{3D_b a^2 v-8D_s a^2 b^2-6D_b b^2-3D_b a^2}{18ab} & \frac{D_s b}{6} & -\frac{D_b(3v-1)}{8} & \cdots \\
\cdots & -\frac{D_s a}{3} & \frac{D_s b}{6} & \frac{D_s\left(b^2+a^2\right)}{3ab} & -\frac{D_s a}{3} & \cdots \\
\cdots & -\frac{3D_b b^2 v-8D_s a^2 b^2-3D_b b^2+12D_b a^2}{36ab} & -\frac{D_b(3v-1)}{8} & -\frac{D_s a}{3} & -\frac{3D_b b^2 v-8D_s a^2 b^2-3D_b b^2-6D_b a^2}{18ab} & \cdots \\
\cdots & -\frac{3D_b a^2 v-8D_s a^2 b^2-6D_b b^2-3D_b a^2}{18ab} & -\frac{D_s b}{3} & -\frac{D_b(3v-1)}{8} & -\frac{3D_b a^2 v-8D_s a^2 b^2+12D_b b^2-3D_b a^2}{36ab} & \cdots \\
\cdots & -\frac{D_s a}{6} & -\frac{D_s b}{6} & -\frac{D_s\left(2b^2-a^2\right)}{6ab} & -\frac{D_s a}{6} & \cdots \\
\cdots & -\frac{D_b(v+1)}{8} & -\frac{D_s a}{6} & \frac{3D_b b^2 v+4D_s a^2 b^2-3D_b b^2+3D_b a^2}{18ab} & -\frac{D_b(3v-1)}{8} & \cdots \\
\cdots & -\frac{D_b(v+1)}{8} & \frac{3D_b a^2 v+4D_s a^2 b^2-6D_b b^2-3D_b a^2}{36ab} & \frac{D_s b}{3} & \frac{D_b(3v-1)}{8} & \cdots
\end{matrix}
\quad (5.26)
$$

The following Module 5.2 shows the entire wxMaxima code for the calculation of the elemental stiffness matrix (geometry: $2a \times 2b \times h$) based on numerical 2×2 integration.

$$
\begin{bmatrix}
\cdots & -\frac{D_s b}{6} & \frac{D_s\left(b^2-2a^2\right)}{6ab} & \frac{D_s a}{3} & -\frac{D_s b}{6} \\
\cdots & \frac{D_b(v+1)}{8} & -\frac{D_s a}{3} & -\frac{3D_b b^2 v-8D_s a^2 b^2-3D_b b^2+12D_b a^2}{36ab} & -\frac{D_b(3v-1)}{8} \\
\cdots & \frac{3D_b a^2 v+4D_s a^2 b^2-6D_b b^2-3D_b a^2}{36ab} & -\frac{D_s b}{6} & \frac{D_b(3v-1)}{8} & \frac{3D_b a^2 v+4D_s a^2 b^2+3D_b b^2-3D_b a^2}{18ab} \\
\cdots & \frac{D_s b}{6} & -\frac{D_s\left(b^2+a^2\right)}{6ab} & \frac{D_s a}{6} & \frac{D_s b}{6} \\
\cdots & \frac{D_b(3v-1)}{8} & -\frac{D_s a}{6} & \frac{3D_b b^2 v+4D_s a^2 b^2-3D_b b^2-6D_b a^2}{36ab} & -\frac{D_b(v+1)}{8} \\
\cdots & \frac{3D_b a^2 v+4D_s a^2 b^2+3D_b b^2-3D_b a^2}{18ab} & -\frac{D_s b}{6} & -\frac{D_b(v+1)}{8} & \frac{3D_b a^2 v+4D_s a^2 b^2-6D_b b^2-3D_b a^2}{36ab} \\
\cdots & \frac{D_s b}{3} & -\frac{D_s\left(2b^2-a^2\right)}{6ab} & -\frac{D_s a}{6} & \frac{D_s b}{3} \\
\cdots & -\frac{D_b(v+1)}{8} & -\frac{D_s a}{6} & \frac{3D_b b^2 v+4D_s a^2 b^2-3D_b b^2+3D_b a^2}{18ab} & \frac{D_b(3v-1)}{8} \\
\cdots & -\frac{3D_b a^2 v-8D_s a^2 b^2-6D_b b^2-3D_b a^2}{18ab} & -\frac{D_s b}{3} & -\frac{D_b(3v-1)}{8} & -\frac{3D_b a^2 v-8D_s a^2 b^2+12D_b b^2-3D_b a^2}{36ab} \\
\cdots & -\frac{D_s b}{3} & \frac{D_s\left(b^2+a^2\right)}{3ab} & -\frac{D_s a}{3} & -\frac{D_s b}{3} \\
\cdots & -\frac{D_b(3v-1)}{8} & -\frac{D_s a}{3} & -\frac{3D_b b^2 v-8D_s a^2 b^2-3D_b b^2-6D_b a^2}{18ab} & \frac{D_b(v+1)}{8} \\
\cdots & -\frac{3D_b a^2 v-8D_s a^2 b^2+12D_b b^2-3D_b a^2}{36ab} & -\frac{D_s b}{3} & \frac{D_b(v+1)}{8} & -\frac{3D_b a^2 v-8D_s a^2 b^2-6D_b b^2-3D_b a^2}{18ab}
\end{bmatrix}
$$

(%i3)
```
kill(all);
load("../Library/my_funs.mac")$
ratprint :false$
fpprintprec : 6$
```

(%o0) *done*

Definition of Timoshenko Plate Stiffness: Nodal Coordinates "ncoor", Material Properties "mat", Thickness Function "thickness", Number of Element Nodes "nen"

(%i7)
```
ncoor : [[0,0],[2*a,0],[2*a,2*b],[0,2*b]]$

mat : [E, v, G, Ks]$

thickness(x,y) := h$

nen : 4$
nqp : 4$
```

Maxima Calculations and Output

(%i12)
```
Ke :
k_el_2D_Reissner_Mindlin_plate_numerical(ncoor, mat, nqp, nen) $

print(" ")$
print("Element stiffness matrix:")$
print(" ")$
print(factor(ratsimp(Ke)))$
```

Element stiffness matrix:

$$\begin{bmatrix}
\frac{D_s\left(b^2+a^2\right)}{3ab} & \frac{D_s a}{3} & \cdots \\
\frac{D_s a}{3} & -\frac{3D_b b^2 v-8D_s a^2 b^2-3D_b b^2-6D_b a^2}{18ab} & \cdots \\
-\frac{D_s b}{3} & -\frac{D_b(v+1)}{8} & \cdots \\
-\frac{D_s\left(2b^2-a^2\right)}{6ab} & \frac{\left(\sqrt{3}-1\right)\left(\sqrt{3}+1\right)D_s a}{12} & \cdots \\
\frac{\left(\sqrt{3}-1\right)\left(\sqrt{3}+1\right)D_s a}{12} & \frac{3D_b b^2 v+4D_s a^2 b^2-3D_b b^2+3D_b a^2}{18ab} & \cdots \\
-\frac{D_s b}{3} & \frac{D_b(3v-1)}{8} & \cdots \\
-\frac{\left(\sqrt{3}-1\right)\left(\sqrt{3}+1\right)D_s\left(b^2+a^2\right)}{12ab} & -\frac{\left(\sqrt{3}-1\right)\left(\sqrt{3}+1\right)D_s a}{12} & \cdots \\
\vdots & \vdots & \vdots
\end{bmatrix}$$

Module 5.2: Calculation of the elemental stiffness matrix based on numerical 2 × 2 integration for the problem shown in Fig. 5.3a (geometry: $2a \times 2b \times h$). The complete stiffness matrix is shown in Eq. (5.26)

Table 5.1 Selected elements of the elemental stiffness matrices

Int.	K_{11}	K_{12}	K_{22}
		Geometry: $2a \times 2b \times h$	
Analyt.	$\frac{D_s(b^2+a^2)}{3ab}$	$\frac{D_s a}{3}$	$-\frac{3D_b b^2 \nu - 8D_s a^2 b^2 - 3D_b b^2 - 6D_b a^2}{18ab}$
1×1	$\frac{D_s(b^2+a^2)}{8ab}$	$\frac{D_s a}{8}$	$-\frac{D_b b^2 \nu - 2D_s a^2 b^2 - D_b b^2 - 2D_b a^2}{16ab}$
2×2	$\frac{D_s(b^2+a^2)}{3ab}$	$\frac{D_s a}{3}$	$-\frac{3D_b b^2 \nu - 8D_s a^2 b^2 - 3D_b b^2 - 6D_b a^2}{18ab}$
3×3	$\frac{D_s(b^2+a^2)}{3ab}$	$\frac{D_s a}{3}$	$-\frac{3D_b b^2 \nu - 8D_s a^2 b^2 - 3D_b b^2 - 6D_b a^2}{18ab}$
		Geometry: $a \times b \times h$	
Analyt.	$\frac{D_s(b^2+a^2)}{3ab}$	$\frac{D_s a}{6}$	$-\frac{3D_b b^2 \nu - 2D_s a^2 b^2 - 3D_b b^2 - 6D_b a^2}{18ab}$
1×1	$\frac{D_s(b^2+a^2)}{8ab}$	$\frac{D_s a}{16}$	$-\frac{2D_b b^2 \nu - D_s a^2 b^2 - 2D_b b^2 - 4D_b a^2}{32ab}$
2×2	$\frac{D_s(b^2+a^2)}{3ab}$	$\frac{D_s a}{6}$	$-\frac{3D_b b^2 \nu - 2D_s a^2 b^2 - 3D_b b^2 - 6D_b a^2}{18ab}$
3×3	$\frac{D_s(b^2+a^2)}{3ab}$	$\frac{D_s a}{6}$	$-\frac{3D_b b^2 \nu - 2D_s a^2 b^2 - 3D_b b^2 - 6D_b a^2}{18ab}$

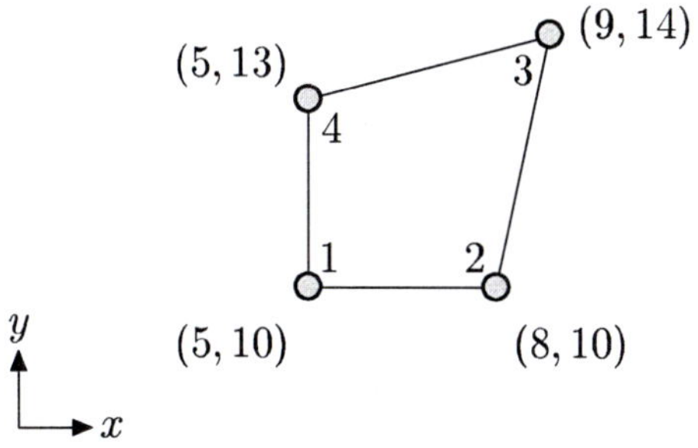

Fig. 5.4 Distorted two-dimensional thick plate element

Table 5.1 summarizes a few elements of the stiffness matrices for the different geometries and integration approaches.

5.2 Example: Distorted two-dimensional thick plate element
Given is a distorted two-dimensional thick plate element as shown in Fig. 5.4. The constant thickness has a value of $h = 1.0$ and the material parameters are given as $E = 70000$ and $\nu = 0.3$. Assume isotropic material behavior. Derive the numerical expression for the elemental stiffness matrix ($k_s = 5/6$). Use analytical, 1×1, 2×2, 3×3, and 4×4 Gauss-Legendre integration. Compare the accuracy of the integration approach for a few selected elements of the stiffness matrix.

5.2 Solution
The following Module 5.3 shows the entire wxMaxima code for the calculation of the elemental stiffness matrix based on analytical integration.

(% i3)
```
kill(all);
load("../Library/my_funs.mac")$
ratprint :false$
fpprintprec : 6$
```

(% o0) *done*

Definition of Timoshenko Plate Stiffness: Nodal Coordinates "ncoor", Material Properties "mat", Thickness Function "thickness", Number of Element Nodes "nen"

(% i14)
```
ncoor : [[5,10],[8,10],[9,14],[5,13]]$

E : 70000$
v : 0.3$
G : E / (2*(1+v))$
Ks : 5/6$
h : 1$

mat : [E, v, G, Ks]$

thickness(x,y) := h$

D_b : E*h^3/(12*(1-v^2))$
D_s : Ks*G*h$

nen : 4$
```

Maxima Calculations and Output

(% i19)
```
Ke :
k_el_2D_Reissner_Mindlin_plate_analytical(ncoor, mat, nen)$

print(" ")$
print("Element stiffness matrix:")$
print(" ")$
print(float(Ke))$
```

Element stiffness matrix:

$$\begin{bmatrix}
14572.7 & 11217.9 & -11217.9 & -4927.41 & 3739.32 & -13087.6 & -4717.9 & \cdots \\
11217.9 & 28985.7 & -626.747 & 7478.63 & 14113.8 & -203.731 & -5608.97 & \cdots \\
-11217.9 & -626.747 & 28985.7 & 13087.6 & -43.4746 & 12030.5 & 5608.97 & \cdots \\
-4927.41 & 7478.63 & 13087.6 & 18899.2 & 14957.3 & 14957.3 & -2956.45 & \cdots \\
3739.32 & 14113.8 & -43.4746 & 14957.3 & 33559.4 & 1487.77 & -11217.9 & \cdots \\
-13087.6 & -203.731 & 12030.5 & 14957.3 & 1487.77 & 33559.4 & 5608.97 & \cdots \\
-4717.9 & -5608.97 & 5608.97 & -2956.45 & -11217.9 & 5608.97 & 10630.8 & \cdots \\
5608.97 & 6568.75 & 873.952 & 14957.3 & 14280.3 & 5.96654 & -11217.9 & \cdots \\
-5608.97 & 873.952 & 6568.75 & 9348.29 & -154.29 & 16363.6 & 11217.9 & \cdots \\
-4927.41 & -13087.6 & -7478.63 & -11015.3 & -7478.63 & -7478.63 & -2956.45 & \cdots \\
13087.6 & 12030.5 & -203.731 & 7478.63 & 5354.25 & -1290.01 & -5608.97 & \cdots \\
-3739.32 & -43.4746 & 14113.8 & 7478.63 & -1290.01 & 5354.25 & 11217.9 & \cdots
\end{bmatrix}$$

Module 5.3: Calculation of the elemental stiffness matrix based on analytical integration for the problem shown in Fig. 5.4. The complete stiffness matrix is shown in Eq. (5.27)

$$
\boldsymbol{K}^{\mathrm{e}} = \begin{bmatrix}
14572.7 & 11217.9 & -11217.9 & -4927.41 & 3739.32 & -13087.6 & -4717.9 & 5608.97 & -5608.97 & -4927.41 & 13087.6 & -3739.32 \\
11217.9 & 28985.7 & -626.747 & 7478.63 & 14113.8 & -203.731 & -5608.97 & 6568.75 & 873.952 & -13087.6 & 12030.5 & -43.4746 \\
-11217.9 & -626.747 & 28985.7 & 13087.6 & -43.4746 & 12030.5 & 5608.97 & 873.952 & 6568.75 & -7478.63 & -203.731 & 14113.8 \\
-4927.41 & 7478.63 & 13087.6 & 18899.2 & 14957.3 & 14957.3 & -2956.45 & 14957.3 & 9348.29 & -11015.3 & 7478.63 & 7478.63 \\
3739.32 & 14113.8 & -43.4746 & 14957.3 & 33559.4 & 1487.77 & -11217.9 & 14280.3 & -154.29 & -7478.63 & 5354.25 & -1290.01 \\
-13087.6 & -203.731 & 12030.5 & 14957.3 & 1487.77 & 33559.4 & 5608.97 & 5.96654 & 16363.6 & -7478.63 & -1290.01 & 5354.25 \\
-4717.9 & -5608.97 & 5608.97 & -2956.45 & -11217.9 & 5608.97 & 10630.8 & -11217.9 & 11217.9 & -2956.45 & -5608.97 & 11217.9 \\
5608.97 & 6568.75 & 873.952 & 14957.3 & 14280.3 & 5.96654 & -11217.9 & 35704.1 & -725.629 & -9348.29 & 16363.6 & -154.29 \\
-5608.97 & 873.952 & 6568.75 & 9348.29 & -154.29 & 16363.6 & 11217.9 & -725.629 & 35704.1 & -14957.3 & 5.96654 & 14280.3 \\
-4927.41 & -13087.6 & -7478.63 & -11015.3 & -7478.63 & -7478.63 & -2956.45 & -9348.29 & -14957.3 & 18899.2 & -14957.3 & -14957.3 \\
13087.6 & 12030.5 & -203.731 & 7478.63 & 5354.25 & -1290.01 & -5608.97 & 16363.6 & 5.96654 & -14957.3 & 33559.4 & 1487.77 \\
-3739.32 & -43.4746 & 14113.8 & 7478.63 & -1290.01 & 5354.25 & 11217.9 & -154.29 & 14280.3 & -14957.3 & 1487.77 & 33559.4
\end{bmatrix} . \tag{5.27}
$$

The following Module 5.4 shows the entire wxMaxima code for the calculation of the elemental stiffness matrix based on numerical 2×2 integration.

Table 5.2 summarizes a few elements of the stiffness matrix for the different integration approaches.

5.2 Elemental and Global Principal Finite Element Equation

5.2.1 Elements with Four Nodes

The principal finite element equation for a single thick four-node plate element (based on the derivations for the stiffness matrix given in Sect. 5.1.1) can be stated as follows (see [2])

$$\begin{bmatrix} K_{11} & K_{12} & K_{13} & \ldots & K_{110} & K_{111} & K_{112} \\ K_{21} & K_{22} & K_{23} & \ldots & K_{210} & K_{211} & K_{212} \\ K_{31} & K_{32} & K_{33} & \ldots & K_{310} & K_{311} & K_{312} \\ \vdots & \vdots & \vdots & \vdots & \vdots & \vdots & \vdots \\ K_{101} & K_{102} & K_{103} & \ldots & K_{1010} & K_{1011} & K_{1012} \\ K_{111} & K_{112} & K_{113} & \ldots & K_{1110} & K_{1111} & K_{1112} \\ K_{121} & K_{122} & K_{123} & \ldots & K_{1210} & K_{1211} & K_{1212} \end{bmatrix} \begin{bmatrix} u_{1z} \\ \phi_{1x} \\ \phi_{1y} \\ \hline u_{2z} \\ \phi_{2x} \\ \phi_{2y} \\ \hline u_{3z} \\ \phi_{3x} \\ \phi_{3y} \\ \hline u_{4z} \\ \phi_{4x} \\ \phi_{4y} \end{bmatrix} = \begin{bmatrix} F_{1z} \\ M_{1x} \\ M_{1y} \\ \hline F_{2z} \\ M_{2x} \\ M_{2y} \\ \hline F_{3z} \\ M_{3x} \\ M_{3y} \\ \hline F_{4z} \\ M_{4x} \\ M_{4y} \end{bmatrix}$$

$$+ \int_A \begin{bmatrix} N_1 & 0 & 0 \\ 0 & N_1 & 0 \\ 0 & 0 & N_1 \\ \hline \vdots & \vdots & \vdots \\ \hline N_4 & 0 & 0 \\ 0 & N_4 & 0 \\ 0 & 0 & N_4 \end{bmatrix} \begin{bmatrix} q_z \\ m_x \\ m_y \end{bmatrix} \mathrm{d}A\,, \tag{5.28}$$

or in abbreviated form

$$\boldsymbol{K}^{\mathrm{e}} \boldsymbol{u}_{\mathrm{p}}^{\mathrm{e}} = \boldsymbol{f}_{t}^{\mathrm{e}} + \boldsymbol{f}_{b}^{\mathrm{e}}\,, \tag{5.29}$$

(% i3)
```
kill(all);
load("../Library/my_funs.mac")$
ratprint :false$
fpprintprec : 6$
```

(% o0) *done*

Definition of Timoshenko Plate Stiffness: Nodal Coordinates "ncoor", Material Properties "mat", Thickness Function "thickness", Number of Element Nodes "nen", Number of Quadrature Points "nqp"

(% i14)
```
ncoor : [[5,10],[8,10],[9,14],[5,13]]$

E : 70000$
v : 0.3$
G : E / (2*(1+v))$
Ks : 5/6$
h : 1$

mat : [E, v, G, Ks]$

thickness(x,y) := h$

D_b : E*h^3/(12*(1-v^2))$
D_s : Ks*G*h$

nen : 4$
nqp : 4$
```

Maxima Calculations and Output

(% i19)
```
Ke :
k_el_2D_Reissner_Mindlin_plate_numerical(ncoor, mat, nqp, nen)$

print(" ")$
print("Element stiffness matrix:")$
print(" ")$
print(float(Ke))$
```

Element stiffness matrix:

$$
\begin{bmatrix}
14546.2 & 11217.9 & -11217.9 & -4906.2 & 3739.32 & -13087.6 & -4733.82 & \cdots \\
11217.9 & 28980.6 & -627.339 & 7478.63 & 14117.9 & -203.257 & -5608.97 & \cdots \\
-11217.9 & -627.339 & 28980.6 & 13087.6 & -43.0002 & 12034.6 & 5608.97 & \cdots \\
-4906.2 & 7478.63 & 13087.6 & 18882.2 & 14957.3 & 14957.3 & -2943.72 & \cdots \\
3739.32 & 14117.9 & -43.0002 & 14957.3 & 33556.1 & 1487.39 & -11217.9 & \cdots \\
-13087.6 & -203.257 & 12034.6 & 14957.3 & 1487.39 & 33556.1 & 5608.97 & \cdots \\
-4733.82 & -5608.97 & 5608.97 & -2943.72 & -11217.9 & 5608.97 & 10621.2 & \cdots \\
5608.97 & 6565.68 & 873.596 & 14957.3 & 14282.7 & 6.25114 & -11217.9 & \cdots \\
-5608.97 & 873.596 & 6565.68 & 9348.29 & -154.005 & 16366.0 & 11217.9 & \cdots \\
-4906.2 & -13087.6 & -7478.63 & -11032.3 & -7478.63 & -7478.63 & -2943.72 & \cdots \\
13087.6 & 12034.6 & -203.257 & 7478.63 & 5350.97 & -1290.39 & -5608.97 & \cdots \\
-3739.32 & -43.0002 & 14117.9 & 7478.63 & -1290.39 & 5350.97 & 11217.9 & \cdots
\end{bmatrix}
$$

Module 5.4: Calculation of the elemental stiffness matrix based on numerical 2×2 integration for the problem shown in Fig. 5.4

Table 5.2 Selected elements of the elemental stiffness matrices for different integration approaches

Int.	K_{11}	K_{12}	K_{22}
Analyt.	14572.7	11217.9	28985.7
1×1	4206.73	4206.73	9224.76
2×2	14546.2	11217.9	28980.6
3×3	14572.6	11217.9	28985.6
4×4	14572.7	11217.9	28985.7

Table 5.3 Post-processing of nodal values for a thick four-node plate element. The distributions are given as being dependent on the nodal values $\boldsymbol{u}_{\mathrm{p}}^{\mathrm{e}}$ as a function of the natural coordinates $-1 \leq \xi \leq +1$ and $-1 \leq \eta \leq +1$. See [2] for further details

Deformation fields

$$\boldsymbol{u}^{\mathrm{e}}(\xi,\eta) = \begin{bmatrix} u_z^{\mathrm{e}}(\xi,\eta) \\ \phi_x^{\mathrm{e}}(\xi,\eta) \\ \phi_y^{\mathrm{e}}(\xi,\eta) \end{bmatrix} = \boldsymbol{N}^{\mathrm{T}}(\xi,\eta)\boldsymbol{u}_{\mathrm{p}}^{\mathrm{e}}$$

General strain fields

$$\boldsymbol{e}(\xi,\eta) = \boldsymbol{\mathcal{L}}_1\boldsymbol{u}^{\mathrm{e}}(\xi,\eta)$$

General stress fields

$$\boldsymbol{s}^*(\xi,\eta) = \boldsymbol{D}^*\boldsymbol{e}(\xi,\eta)$$

where $\boldsymbol{K}^{\mathrm{e}}$ is the elemental stiffness matrix (the calculation is to be performed according to the scheme given on Sect. 4.1.1), $\boldsymbol{u}_{\mathrm{p}}^{\mathrm{e}}$ is the column matrix of nodal unknowns, $\boldsymbol{f}_t^{\mathrm{e}}$ is the boundary force column matrix, and $\boldsymbol{f}_b^{\mathrm{e}}$ is body force column matrix. The four[4] interpolation functions N_i are given in Eqs. (5.12)–(5.15) for an element in the unit space.

Once the nodal deformations are known, e.g. based on $\boldsymbol{u}_{\mathrm{p}}^{\mathrm{e}} = (\boldsymbol{K}^{\mathrm{e}})^{-1}\boldsymbol{f}^{\mathrm{e}}$, further quantities can be calculated based on this result (the so-called post-processing, see Table 5.3).

[4] Under the assumption $N_{\phi_y} = N_{\phi_x} = N_u$.

5.2.2 *Maxima Modules for the Finite Element Equation*

5.2.3 *Maxima Examples for the Principal Finite Element Equation*

5.3 Example: One-element example of a cantilever thick plate

Given is a cantilever thick plate element as indicated in Fig. 5.5. The side lengths are equal to $2a$. The plate is loaded by two single forces $\frac{1}{2}F_0$ acting at the right-hand nodes of the element. The material is described based on the engineering constants Young's modulus E and Poisson's ratio ν. Assume isotropic material behavior.

Use a single thick plate element in the following to model the problem and to calculate the nodal unknowns based on analytical and numerical integration. Compare your results with the solution for a classical plate element.

5.3 Solution
The following Module 5.5 shows the entire wxMaxima code for the calculation of the nodal unknowns based on analytical integration.

The following Module 5.6 shows the entire wxMaxima code for the calculation of the nodal unknowns based on numerical 2×2 integration.

The comparison between the thin (see Example 4.3) and the here presented thick plate solution is given in Table 5.4.

5.4 Example: Four-element example of a thick plate fixed at all four edges
Given is a thick plate which is fixed at all four sides, see Fig. 5.6. The side lengths are $4a \times 4b$. The plate is loaded by a single forces F_0 acting in the middle of the plate. The material is described based on the engineering constants Young's modulus E and Poisson's ratio ν. Assume isotropic material behavior.

Use four thick plate elements (each $2a \times 2b \times h$) in the following to model the problem and to calculate the nodal unknowns in the middle of the plate based on analytical and numerical integration.

5.4 Solution
The following Module 5.7 shows the entire wxMaxima code for the calculation of the nodal unknowns based on analytical integration.

Fig. 5.5 Cantilever square thick plate element

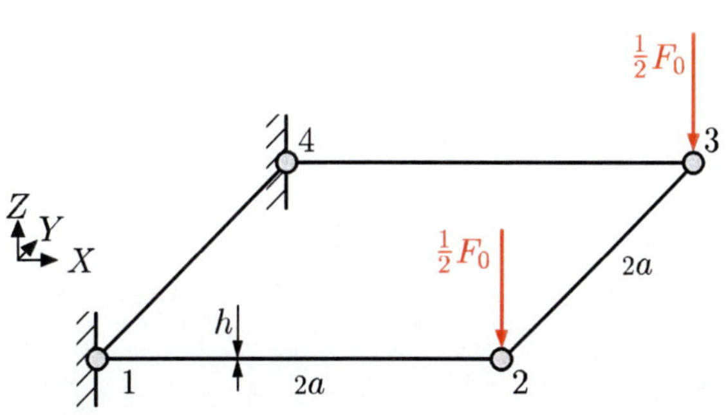

(%i3)
```
kill(all);
load("../Library/my_funs.mac")$
ratprint :false$
fpprintprec : 6$
```

(%o0) *done*

Definition of Reissner-Mindlin Plate Problem: Nodal Coordinates "nodxy", Element Connectivity "elenod", Element Material Properties "elemat", Number of Element Nodes "nen", Thickness Function "thickness", freedom tags list "nodtag", freedom values list "nodval", Numerical or Analytical Calculations Switch "numerical"

(%i14)
```
nodxy : [[-a,-a],[a,-a],[a,a],[-a,a]]$

elenod : [[1,2,3,4]]$

elemat : [[E, v]]$

thickness(x,y) := h$

/* body forces */
q_z(x,y) := 0$
m_x(x,y) := 0$
m_y(x,y) := 0$

/*
D_b : E*h^3/(12*(1-v^2))$
D_s : Ks*G*h$
*/

nodtag : [[1,1,1],[0,0,0],[0,0,0],[1,1,1]]$
nodval : [[0,0,0],[-F_0/2,0,0],[-F_0/2,0,0],[0,0,0]]$

numerical : false$

nen : 4$
```

Maxima Calculations and Output

(%i16)
```
[noddis,Kmod,fmod] :
ReissnerMindlinPlateElementsSolution(nodxy,elenod,elemat,nodtag,nodval,
                                     numerical)$

for i : 1 thru length(nodxy) do (
  print(" "),
  print("Displacements at Node ", i),
  print(" ", U_Z, "=", factor(ratsimp(noddis[i][1]))),
  print(" ", φ_X, "=", factor(ratsimp(noddis[i][2]))),
  print(" ", φ_Y, "=", factor(ratsimp(noddis[i][3])))
)$
```

Displacements at Node 1

$$U_Z = 0$$

$$\varphi_X = 0$$

$$\varphi_Y = 0$$

Displacements at Node 2

$$U_Z = \frac{F_0\left(54D_b{}^2v^2 + 12D_bD_sa^2v + 9D_b{}^2v - 32D_s{}^2a^4 - 132D_bD_sa^2 - 81D_b{}^2\right)}{D_s\left(54D_b{}^2v^2 + 3D_bD_sa^2v + 9D_b{}^2v - 8D_s{}^2a^4 - 51D_bD_sa^2 - 81D_b{}^2\right)}$$

$$\varphi_X = \frac{54D_bF_0av}{54D_b{}^2v^2 + 3D_bD_sa^2v + 9D_b{}^2v - 8D_s{}^2a^4 - 51D_bD_sa^2 - 81D_b{}^2}$$

$$\varphi_Y = \frac{54D_bF_0av}{54D_b{}^2v^2 + 3D_bD_sa^2v + 9D_b{}^2v - 8D_s{}^2a^4 - 51D_bD_sa^2 - 81D_b{}^2}$$

Displacements at Node 3

$$U_Z = \frac{F_0\left(54D_b{}^2v^2 + 12D_bD_sa^2v + 9D_b{}^2v - 32D_s{}^2a^4 - 132D_bD_sa^2 - 81D_b{}^2\right)}{D_s\left(54D_b{}^2v^2 + 3D_bD_sa^2v + 9D_b{}^2v - 8D_s{}^2a^4 - 51D_bD_sa^2 - 81D_b{}^2\right)}$$

$$\varphi_X = -\frac{54D_bF_0av}{54D_b{}^2v^2 + 3D_bD_sa^2v + 9D_b{}^2v - 8D_s{}^2a^4 - 51D_bD_sa^2 - 81D_b{}^2}$$

$$\varphi_Y = -\frac{54D_bF_0av}{54D_b{}^2v^2 + 3D_bD_sa^2v + 9D_b{}^2v - 8D_s{}^2a^4 - 51D_bD_sa^2 - 81D_b{}^2}$$

Displacements at Node 4

$$U_Z = 0$$
$$\varphi_X = 0$$
$$\varphi_Y = 0$$

Module 5.5: Calculation of the nodal unknowns based on analytical integration for the problem shown in Fig. 5.5

The following Module 5.8 shows the entire wxMaxima code for the calculation of the nodal unknowns based on numerical 2×2 integration.

5.5 Example: Four-element example of a thick plate fixed at all four edges
Given is a thick plate which is fixed at all four sides, see Fig. 5.7. The side lengths are equal to $4a$. The plate is loaded by a single forces F_0 acting in the middle of the

Table 5.4 Comparison between the thin and the thick plate solution for analytical integration: $E = 210000$, $\nu = 0.3$, $h = 1.0$, $a = 4.0$, $F_0 = 1000$, $k_s = 5/6$ (consistent units and isotropic material assumed)

DOF	Thin	Thick
u_Z	−1.129142857142857	−0.05716957921079345
φ_X or ϕ_X	−0.02476190476190476	$-1.204846848289987 \times 10^{-4}$
φ_Y or ϕ_Y	−0.02476190476190476	$-1.204846848289987 \times 10^{-4}$

(% i3)
```
kill(all);
load("../Library/my_funs.mac")$
ratprint :false$
fpprintprec : 6$
```

(% o0) *done*

Definition of Reissner-Mindlin Plate Problem: Nodal Coordinates "nodxy", Element Connectivity "elenod", Element Material Properties "elemat", Number of Element Nodes "nen", Thickness Function "thickness", freedom tags list "nodtag", freedom values list "nodval", Numerical or Analytical Calculations Switch "numerical"

(% i14)
```
nodxy : [[-a,-a],[a,-a],[a,a],[-a,a]]$

elenod : [[1,2,3,4]]$

elemat : [[E, v]]$

thickness(x,y) := h$

/* body forces */
q_z(x,y) := 0$
m_x(x,y) := 0$
m_y(x,y) := 0$

/*
D_b : E*h^3/(12*(1-v^2))$
D_s : Ks*G*h$
*/

nodtag : [[1,1,1],[0,0,0],[0,0,0],[1,1,1]]$
nodval : [[0,0,0],[-F_0/2,0,0],[-F_0/2,0,0],[0,0,0]]$

numerical : true$

nen : 4$
nqp : 4$
```

Maxima Calculations and Output

(% i18)
```
[noddis,Kmod,fmod] :
ReissnerMindlinPlateElementsSolution(nodxy,elenod,elemat,nodtag,nodval,
                                                        numerical)$

coeff : mat[1]*(th^3)/(12*(1-mat[2]^2))$
noddis : noddis*coeff/D$

for i : 1 thru length(nodxy) do (
  print(" "),
  print("Displacements at Node ", i),
  print(" ", U_Z, "=", factor(ratsimp(noddis[i][1]))),
  print(" ", φ_X, "=", factor(ratsimp(noddis[i][2]))),
  print(" ", φ_Y, "=", factor(ratsimp(noddis[i][3])))
)$
```

Displacements at Node 1

$$U_Z = 0$$
$$\varphi_X = 0$$
$$\varphi_Y = 0$$

Displacements at Node 2

$$U_Z = -\frac{F_0\left(54{D_\mathrm{b}}^2v^2 + 12D_\mathrm{b}D_\mathrm{s}a^2v + 9{D_\mathrm{b}}^2v - 32{D_\mathrm{s}}^2a^4 - 132D_\mathrm{b}D_\mathrm{s}a^2 - 81{D_\mathrm{b}}^2\right)}{D_\mathrm{s}\left(54{D_\mathrm{b}}^2v^2 + 3D_\mathrm{b}D_\mathrm{s}a^2v + 9{D_\mathrm{b}}^2v - 8{D_\mathrm{s}}^2a^4 - 51D_\mathrm{b}D_\mathrm{s}a^2 - 81{D_\mathrm{b}}^2\right)}$$

$$\varphi_X = \frac{54D_\mathrm{b}F_0av}{54{D_\mathrm{b}}^2v^2 + 3D_\mathrm{b}D_\mathrm{s}a^2v + 9{D_\mathrm{b}}^2v - 8{D_\mathrm{s}}^2a^4 - 51D_\mathrm{b}D_\mathrm{s}a^2 - 81{D_\mathrm{b}}^2}$$

$$\varphi_Y = \frac{3F_0a\left(3D_\mathrm{b}v - 8D_\mathrm{s}a^2 - 27D_\mathrm{b}\right)}{54{D_\mathrm{b}}^2v^2 + 3D_\mathrm{b}D_\mathrm{s}a^2v + 9{D_\mathrm{b}}^2v - 8{D_\mathrm{s}}^2a^4 - 51D_\mathrm{b}D_\mathrm{s}a^2 - 81{D_\mathrm{b}}^2}$$

Displacements at Node 3

$$U_Z = -\frac{F_0\left(54{D_\mathrm{b}}^2v^2 + 12D_\mathrm{b}D_\mathrm{s}a^2v + 9{D_\mathrm{b}}^2v - 32{D_\mathrm{s}}^2a^4 - 132D_\mathrm{b}D_\mathrm{s}a^2 - 81{D_\mathrm{b}}^2\right)}{D_\mathrm{s}\left(54{D_\mathrm{b}}^2v^2 + 3D_\mathrm{b}D_\mathrm{s}a^2v + 9{D_\mathrm{b}}^2v - 8{D_\mathrm{s}}^2a^4 - 51D_\mathrm{b}D_\mathrm{s}a^2 - 81{D_\mathrm{b}}^2\right)}$$

$$\varphi_X = -\frac{54D_\mathrm{b}F_0av}{54{D_\mathrm{b}}^2v^2 + 3D_\mathrm{b}D_\mathrm{s}a^2v + 9{D_\mathrm{b}}^2v - 8{D_\mathrm{s}}^2a^4 - 51D_\mathrm{b}D_\mathrm{s}a^2 - 81{D_\mathrm{b}}^2}$$

$$\varphi_Y = \frac{3F_0a\left(3D_\mathrm{b}v - 8D_\mathrm{s}a^2 - 27D_\mathrm{b}\right)}{54{D_\mathrm{b}}^2v^2 + 3D_\mathrm{b}D_\mathrm{s}a^2v + 9{D_\mathrm{b}}^2v - 8{D_\mathrm{s}}^2a^4 - 51D_\mathrm{b}D_\mathrm{s}a^2 - 81{D_\mathrm{b}}^2}$$

Displacements at Node 4

$$U_Z = 0$$
$$\varphi_X = 0$$
$$\varphi_Y = 0$$

Module 5.6: Calculation of the nodal unknowns based on numerical 2×2 integration for the problem shown in Fig. 5.5

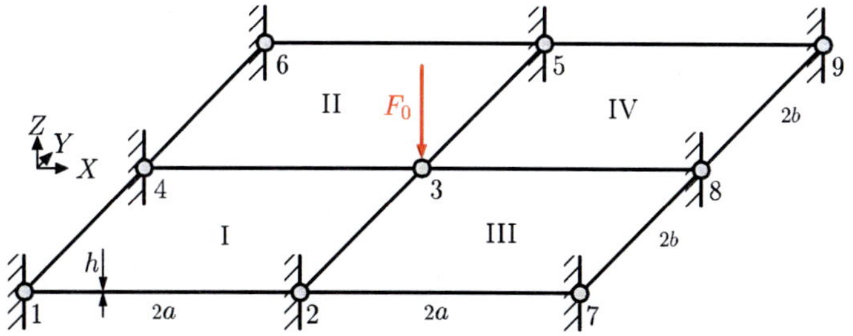

Fig. 5.6 Thick plate problem ($4a \times 4b$) with four edges fixed

```
(% i3)  kill(all);
        load("../Library/my_funs.mac")$
        ratprint :false$
        fpprintprec : 6$

(% o0)  done
```

Definition of Reissner-Mindlin Plate Problem: Nodal Coordinates "nodxy", Element Connectivity "elenod", Element Material Properties "elemat", Number of Element Nodes "nen", Thickness Function "thickness", freedom tags list "nodtag", freedom values list "nodval", Numerical or Analytical Calculations Switch "numerical"

```
(% i16) nodxy : [[0,0],[2*a,0],[2*a,2*b],[0,2*b],[2*a,4*b],[0,4*b],[4*a,0],[4*a,2*b],[4*a,4*b]]$

        elenod : [[1,2,3,4],[4,3,5,6],[2,7,8,3],[3,8,9,5]]$

        elemat : makelist([E, v],4)$

        thickness(x,y) := h$

        /* body forces */
        q_z(x,y) := 0$
        m_x(x,y) := 0$
        m_y(x,y) := 0$

        /*
        D_b : E*h^3/(12*(1-v^2))$
        D_s : Ks*G*h$
        */
        nodtag : makelist([1,1,1],9)$
        nodtag[3] : [0,0,0]$
        nodval : makelist([0,0,0],9)$
        nodval[3] : [-F_0,0,0]$

        numerical : false$
        nen : 4$
```

Maxima Calculations and Output

```
(% i16) [noddis,Kmod,fmod] :
        ReissnerMindlinPlateElementsSolution(nodxy,elenod,elemat,nodtag,nodval,
                                                                 numerical)$

        print(" ")$
        print("Displacements at Node ", 3)$
        print(" ", U_Z, "=", factor(ratsimp(noddis[3][1])))$
        print(" ", φ_X, "=", factor(ratsimp(noddis[3][2])))$
        print(" ", φ_Y, "=", factor(ratsimp(noddis[3][3])))$
```

Displacements at Node 3

$$U_Z = -\frac{3F_0ab}{4D_s\left(b^2+a^2\right)}$$

$$\varphi_X = 0$$

$$\varphi_Y = 0$$

Module 5.7: Calculation of the nodal unknowns based on analytical integration for the problem shown in Fig. 5.6

(% i3)
```
kill(all);
load("../Library/my_funs.mac")$
ratprint :false$
fpprintprec : 6$
```

(% o0) *done*

Definition of Reissner-Mindlin Plate Problem: Nodal Coordinates "nodxy", Element Connectivity "elenod", Element Material Properties "elemat", Number of Element Nodes "nen", Thickness Function "thickness", freedom tags list "nodtag", freedom values list "nodval", Numerical or Analytical Calculations Switch "numerical"

(% i16)
```
nodxy : [[0,0],[2*a,0],[2*a,2*b],[0,2*b],[2*a,4*b],[0,4*b],[4*a,0],[4*a,2*b],[4*a,4*b]]$

elenod : [[1,2,3,4],[4,3,5,6],[2,7,8,3],[3,8,9,5]]$

elemat : makelist([E, v],4)$

thickness(x,y) := h$

/* body forces */
q_z(x,y) := 0$
m_x(x,y) := 0$
m_y(x,y) := 0$

/*
D_b : E*h^3/(12*(1-v^2))$
D_s : Ks*G*h$
*/
nodtag : makelist([1,1,1],9)$
nodtag[3] : [0,0,0]$
nodval : makelist([0,0,0],9)$
nodval[3] : [-F_0,0,0]$

numerical : true$
nen : 4$
nqp : 4$
```

Maxima Calculations and Output

(% i16)
```
[noddis,Kmod,fmod] :
ReissnerMindlinPlateElementsSolution(nodxy,elenod,elemat,nodtag,nodval,
                                      numerical)$

print(" ")$
print("Displacements at Node ", 3)$
print(" ", U_Z, "=", factor(ratsimp(noddis[3][1])))$
print(" ", φ_X, "=", factor(ratsimp(noddis[3][2])))$
print(" ", φ_Y, "=", factor(ratsimp(noddis[3][3])))$
```

Displacements at Node 3

$$U_Z = -\frac{3F_0ab}{4D_S\left(b^2+a^2\right)}$$

$$\varphi_X = 0$$

$$\varphi_Y = 0$$

Module 5.8: Calculation of the nodal unknowns based on numerical 2×2 integration for the problem shown in Fig. 5.6

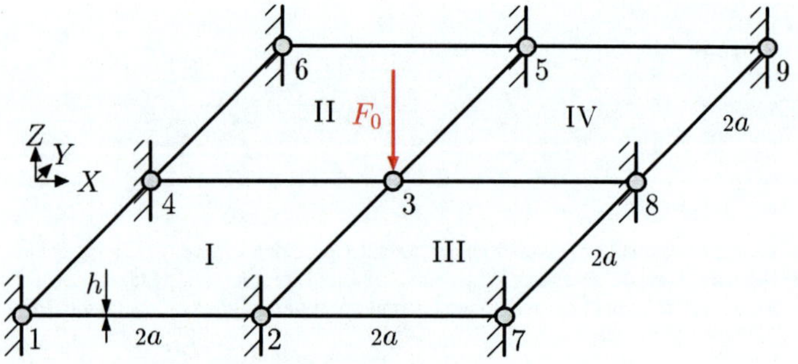

Fig. 5.7 Thick plate problem with four edges fixed

plate. The material is described based on the engineering constants Young's modulus E and Poisson's ratio ν. Assume isotropic material behavior.

Use four thick plate elements (each $2a \times 2a \times h$) in the following to model the problem and to calculate the nodal unknowns in the middle of the plate based on analytical and numerical integration. Compare your results with the solution for a classical plate element.

5.5 Solution

The following Module 5.9 shows the entire wxMaxima code for the calculation of the nodal unknowns based on analytical integration.

The following Module 5.10 shows the entire wxMaxima code for the calculation of the nodal unknowns based on numerical 2×2 integration.

The following Module 5.11 shows the entire wxMaxima code for the calculation of the nodal unknowns based on numerical 1×1 integration.

The comparison between the thin (see Example 4.4) and the here presented thick plate solution is given in Table 5.5.

```
(% i3)  kill(all);
        load("../Library/my_funs.mac")$
        ratprint :false$
        fpprintprec : 6$

(% o0)  done
```

Definition of Reissner-Mindlin Plate Problem: Nodal Coordinates "nodxy", Element Connectivity "elenod", Element Material Properties "elemat", Number of Element Nodes "nen", Thickness Function "thickness", freedom tags list "nodtag", freedom values list "nodval", Numerical or Analytical Calculations Switch "numerical"

```
(% i16) nodxy : [[0,0],[2*a,0],[2*a,2*a],[0,2*a],[2*a,4*a],[0,4*a],[4*a,0],[4*a,2*a],[4*a,4*a]]$

        elenod : [[1,2,3,4],[4,3,5,6],[2,7,8,3],[3,8,9,5]]$

        elemat : makelist([E, v],4)$

        thickness(x,y) := h$

        /* body forces */
        q_z(x,y) := 0$
        m_x(x,y) := 0$
        m_y(x,y) := 0$

        /*
         D_b : E*h^3/(12*(1-v^2))$
         D_s : Ks*G*h$
        */
        nodtag : makelist([1,1,1],9)$
        nodtag[3] : [0,0,0]$
        nodval : makelist([0,0,0],9)$
        nodval[3] : [-F_0,0,0]$

        numerical : false$
        nen : 4$
```

Maxima Calculations and Output

```
(% i16) [noddis,Kmod,fmod] :
        ReissnerMindlinPlateElementsSolution(nodxy,elenod,elemat,nodtag,nodval,
                                             numerical)$

        print(" ")$
        print("Displacements at Node ", 3)$
        print(" ", U_Z, "=", factor(ratsimp(noddis[3][1])))$
        print(" ", φ_X, "=", factor(ratsimp(noddis[3][2])))$
        print(" ", φ_Y, "=", factor(ratsimp(noddis[3][3])))$
```

Displacements at Node 3

$$U_Z = -\frac{3F_0}{8D_s}$$

$$\varphi_X = 0$$

$$\varphi_Y = 0$$

Module 5.9: Calculation of the nodal unknowns based on analytical integration for the problem shown in Fig. 5.7

(%i3)
```
kill(all);
load("../Library/my_funs.mac")$
ratprint :false$
fpprintprec : 6$
```

(%o0) *done*

Definition of Reissner-Mindlin Plate Problem: Nodal Coordinates "nodxy", Element Connectivity "elenod", Element Material Properties "elemat", Number of Element Nodes "nen", Thickness Function "thickness", freedom tags list "nodtag", freedom values list "nodval", Numerical or Analytical Calculations Switch "numerical"

(%i16)
```
nodxy : [[0,0],[2*a,0],[2*a,2*a],[0,2*a],[2*a,4*a],[0,4*a],[4*a,0],[4*a,2*a],[4*a,4*a]]$

elenod : [[1,2,3,4],[4,3,5,6],[2,7,8,3],[3,8,9,5]]$

elemat : makelist([E, v],4)$

thickness(x,y) := h$

/* body forces */
q_z(x,y) := 0$
m_x(x,y) := 0$
m_y(x,y) := 0$

/*
D_b : E*h^3/(12*(1-v^2))$
D_s : Ks*G*h$
*/
nodtag : makelist([1,1,1],9)$
nodtag[3] : [0,0,0]$
nodval : makelist([0,0,0],9)$
nodval[3] : [-F_0,0,0]$

numerical : true$
nen : 4$
nqp : 4$
```

Maxima Calculations and Output

(%i16)
```
[noddis,Kmod,fmod] :
ReissnerMindlinPlateElementsSolution(nodxy,elenod,elemat,nodtag,nodval,
                                     numerical)$

print(" ")$
print("Displacements at Node ", 3)$
print(" ", U_Z, "=", factor(ratsimp(noddis[3][1])))$
print(" ", φ_X, "=", factor(ratsimp(noddis[3][2])))$
print(" ", φ_Y, "=", factor(ratsimp(noddis[3][3])))$
```

Displacements at Node 3

$$U_Z = -\frac{3F_0}{8D_s}$$

$$\varphi_X = 0$$

$$\varphi_Y = 0$$

Module 5.10: Calculation of the nodal unknowns based on numerical 2×2 integration for the problem shown in Fig. 5.7

```
(%i3)  kill(all);
       load("../Library/my_funs.mac")$
       ratprint :false$
       fpprintprec : 6$

(%o0)  done
```

Definition of Reissner-Mindlin Plate Problem: Nodal Coordinates "nodxy", Element Connectivity "elenod", Element Material Properties "elemat", Number of Element Nodes "nen", Thickness Function "thickness", freedom tags list "nodtag", freedom values list "nodval", Numerical or Analytical Calculations Switch "numerical"

```
(%i16) nodxy : [[0,0],[2*a,0],[2*a,2*a],[0,2*a],[2*a,4*a],[0,4*a],[4*a,0],[4*a,2*a],[4*a,4*a]]$

       elenod : [[1,2,3,4],[4,3,5,6],[2,7,8,3],[3,8,9,5]]$

       elemat : makelist([E, v],4)$

       thickness(x,y) := h$

       /* body forces */
       q_z(x,y) := 0$
       m_x(x,y) := 0$
       m_y(x,y) := 0$

       /*
       D_b : E*h^3/(12*(1-v^2))$
       D_s : Ks*G*h$
       */
       nodtag : makelist([1,1,1],9)$
       nodtag[3] : [0,0,0]$
       nodval : makelist([0,0,0],9)$
       nodval[3] : [-F_0,0,0]$

       numerical : true$
       nen : 4$
       nqp : 1$
```

Maxima Calculations and Output

```
(%i16) [noddis,Kmod,fmod] :
       ReissnerMindlinPlateElementsSolution(nodxy,elenod,elemat,nodtag,nodval,
                                            numerical)$

       print(" ")$
       print("Displacements at Node ", 3)$
       print(" ", U_Z, "=", factor(ratsimp(noddis[3][1])))$
       print(" ", φ_X, "=", factor(ratsimp(noddis[3][2])))$
       print(" ", φ_Y, "=", factor(ratsimp(noddis[3][3])))$
```

Displacements at Node 3

$$U_Z = -\frac{F_0}{2D_s}$$

$$\varphi_X = 0$$

$$\varphi_Y = 0$$

Module 5.11: Calculation of the nodal unknowns based on numerical 1×1 integration for the problem shown in Fig. 5.7

Table 5.5 Comparison between the thin and the thick plate solution for different integration techniques: $E = 70000$, $\nu = 0.3$, $h = 1.0$, $a = 4.0$, $k_s = 5/6$ (consistent units and isotropic material assumed)

DOF	Thin	Thick
	Analytical integration	
u_Z	$-1.47727 \times 10^{-5} \times F_0$	$-1.67143 \times 10^{-5} \times F_0$
φ_X or ϕ_X	0	0
φ_Y or ϕ_Y	0	0
	Numerical 2×2 integration	
u_Z	$-1.56 \times 10^{-5} \times F_0$	$-1.67143 \times 10^{-5} \times F_0$
φ_X or ϕ_X	0	0
φ_Y or ϕ_Y	0	0
	Numerical 1×1 integration	
u_Z	$-5.57143 \times 10^{-5} \times F_0$	$-2.22857 \times 10^{-5} \times F_0$
φ_X or ϕ_X	0	0
φ_Y or ϕ_Y	0	0

References

1. Mindlin RD (1951) Influence of rotary inertia and shear on flexural motions isotropic, elastic plates. J Appl Mech-T ASME 18:1031–1036
2. Öchsner A (2020) Computational statics and dynamics - an introduction based on the finite element method. Springer, Singapore
3. Reissner E (1945) The effect of transverse shear deformation on the bending of elastic plates. J Appl Mech-T ASME 12:A68–A77

Chapter 6
Graphical User Interface

The graphical user interface (GUI) allows a different way of defining the model definition compared to the previous chapters, which were rather based on a command-line approach. Providing the definition of a model directly in wxMaxima allows for much more flexibility but the data input approach is not very comfortable. Thus, the GUI allows this comfortable data input but not all command-cbased options as in wxMaxima are available. The GUI is written in the programming environment wxPython [1] under consideration of particular libraries, see Sect. 7.2 for details of the source code.

The application of the GUI is explained in the following based on the previous example Sect. 5.5, see Sect. 5.2.3 for more details.

After executing the Python command (e.g. `python3 main.py`), the initial window as shown in Fig. 6.1 appears. This window allows the basic definition of the element type ('Problem description' tab), i.e., plane elasticity (either as plane stress or plane strain formulation), classical plate, or shear deformable plate element. In addition to the computational approach, i.e., analytical or numerical integration, the user can load or save the model under a particular name. This allows to open a previous model and to modify the definitions. Thus, parametric studies can be easily performed.

Once a particular problem type is chosen, the window provides further options to define the material and geometrical properties, as well as further information on the number of nodes and elements, and body forces, see Fig. 6.2.

Selecting the 'Nodal Coordinates and Element Connectivity' tab allows to provide the coordinates of each node and to define each single element based on its corner nodes, see Fig. 6.3.

Selecting the 'Boundary Conditions' tab allows to define the constraints or loads at each single node, see Fig. 6.4.

Selecting the 'Output' tab allows to define the output mode, see Fig. 6.5. The user can request to export a classical wxMaxima file or to directly request wxMaxima to execute the calculation.

A. Öchsner and R. Makvandi, *Plane Finite Elements for Two-Dimensional Problems*,
https://doi.org/10.1007/978-3-030-89550-1_6

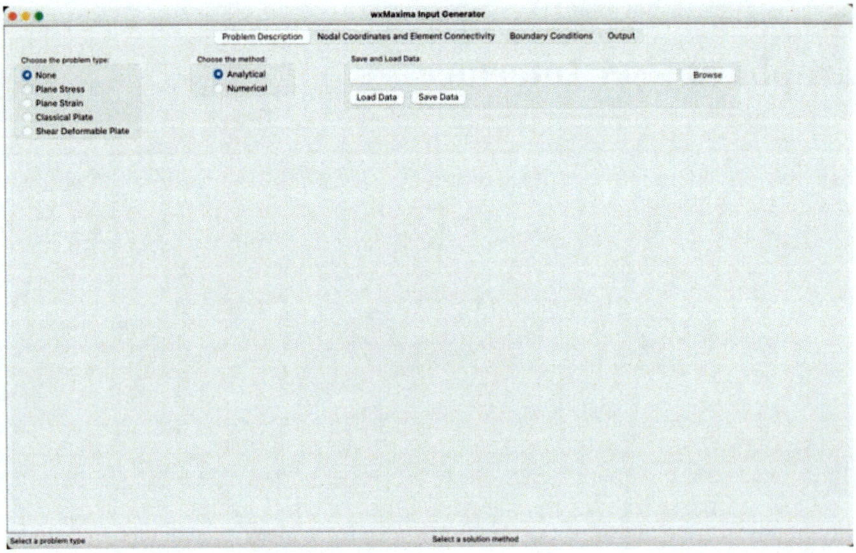

Fig. 6.1 Initial window (Problem description) of the GUI

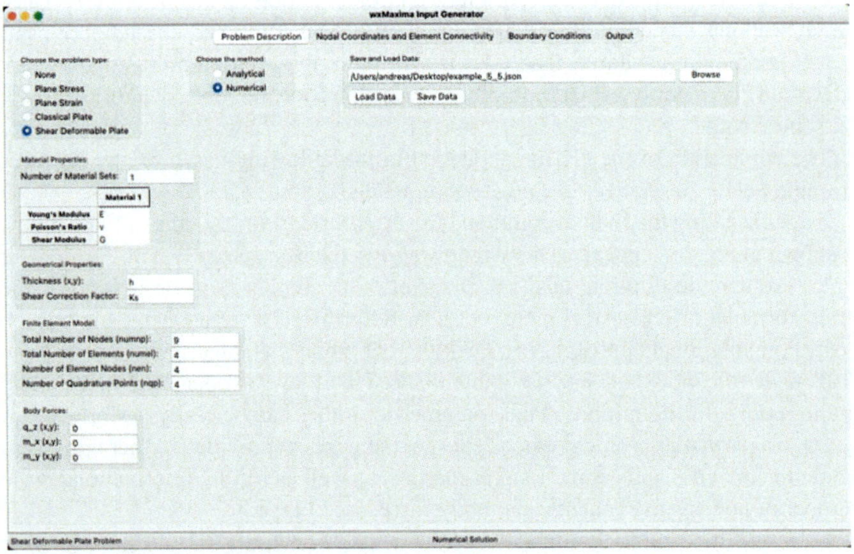

Fig. 6.2 Example of the definition of a shear deformable plate

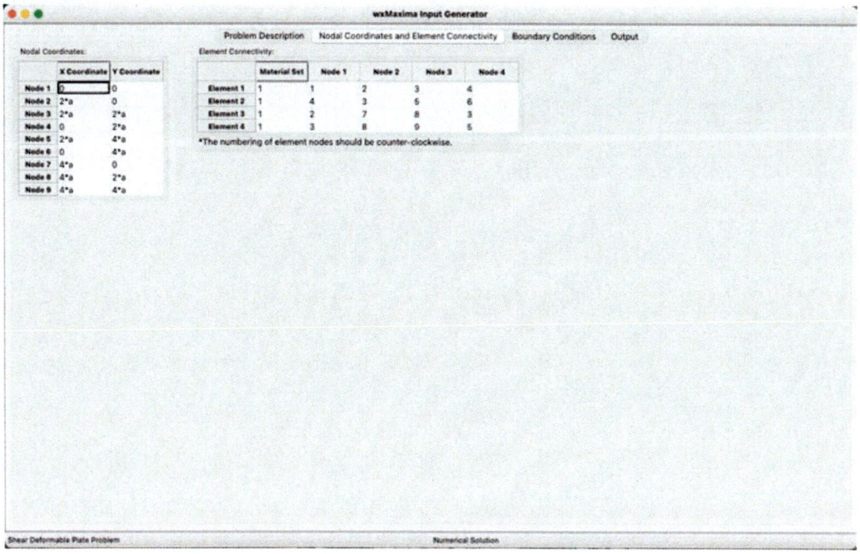

Fig. 6.3 Definition of nodal coordinates and element connectivity in the GUI

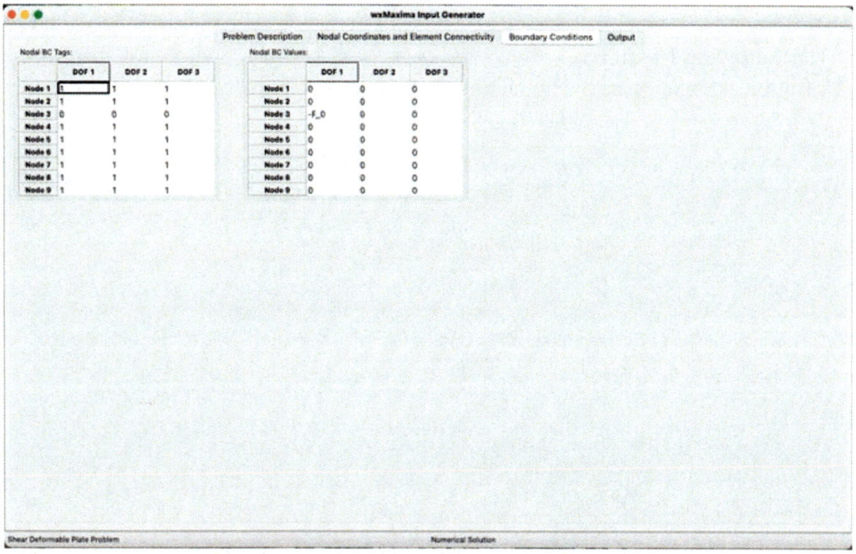

Fig. 6.4 Definition of the boundary conditions in the GUI

Fig. 6.5 Output definition in the GUI

The following Module 6.1 shows the entire wxMaxima code obtained from the GUI for the calculation of the nodal unknowns based on numerical 2×2 integration.

```
kill(all);
load("../Library/my_funs.mac")$
fpprintprec : 6$ /* float numbers precision (only for printing) */
ratprint : false$
```

Definition of Shear Deformable Plate Elasticity Problem: Nodal Coordinates "nodxy", Element Connectivity "elenod", Element Material Properties "elemat", Number of Quadrature Points "nqp", Number of Element Nodes "nen", Thickness Function "thickness", freedom tags list "nodtag", freedom values list "nodval", Numerical or Analytical Calculations Switch "numerical"

```
nodxy : [[0,0],[2*a,0],[2*a,2*a],[0,2*a],[2*a,4*a],[0,4*a],[4*a,0],[4*a,2*a],[4*a,4*a]]$
elenod : [[1,2,3,4],[4,3,5,6],[2,7,8,3],[3,8,9,5]]$
elemat : [[E, v,G], [E, v,G], [E, v,G], [E, v,G]]$
numnod : length(nodxy)$
numele : length(elenod)$

thickness(x,y) := h$

q_z(x,y) := 0$
m_x(x,y) := 0$
m_y(x,y) := 0$
```

```
/*
/D_b : 0*(h)^3/(12*(1-0^2))$
/D_s : (Ks)*(0)*(h)$
*/

nodtag : [[1, 1, 1], [1, 1, 1], [0, 0, 0], [1, 1, 1], [1, 1, 1], [1, 1, 1], [1, 1, 1], [1, 1, 1],
          [1, 1, 1]]$
nodval : [[0, 0, 0], [0, 0, 0], [-F_0, 0, 0], [0, 0, 0], [0, 0, 0], [0, 0, 0], [0, 0, 0],
          [0, 0, 0],[0, 0, 0]]$

nen : 4$
numerical : true$
nqp : 4$

Maxima Calculations and Output

-->    [noddis,Kmod,fmod] :
       ReissnerMindlinPlateElementsSolution(nodxy,elenod,elemat,nodtag,nodval,
                                            numerical)$

       for i : 1 thru length(nodxy) do(
          print(" ")$
          print("Displacements at Node ", i)$
          print(" ", U_Z, "=", factor(ratsimp(noddis[i][1])))$
          print(" ", φ_X, "=", factor(ratsimp(noddis[i][2])))$
          print(" ", φ_Y, "=", factor(ratsimp(noddis[i][2])))$
       )$
```

Module 6.1: Maxima file produced by the GUI for the problem shown in Fig. 5.7

Reference

1. wxPython - The GUI toolkit for Python. https://wxpython.org. Accessed 20 Mar 2021

Chapter 7
Source Codes

7.1 Maxima Library

The following file my_funs.mac must be included in all Maxima sheets for correct execution and result display.

To easier navigate in the source code of the file my_funs.mac, Table 7.1 collects in alphabetical order all the functions and the corresponding line in the source code.

my_funs.mac (main file)

```
/*
load("../Library/my_funs.mac");
*/


/*
2D linear elasticity element
Classical Plate Elements
elemental force matrix
analytical integration

Inputs:
ncoor(2,4)  : nodal point coordinates
          ncoor[1,:] : x-direction
          ncoor[2,:] : y-direction

nen         : number of nodes per element

body        : body forces in the x- and y-directions

Output:

```

Supplementary Information The online version contains supplementary material available at https://doi.org/10.1007/978-3-030-89550-1_7.

A. Öchsner and R. Makvandi, *Plane Finite Elements for Two-Dimensional Problems*,
https://doi.org/10.1007/978-3-030-89550-1_7

Table 7.1 Alphabetical list of all functions in the main library `my_funs.mac`

Function name	Program line
body_2D_plane_analytical(ncoor, nen, body)	25
body_el_2D_Kirschhoff_plate_analytical(ncoor, nen)	88
body_el_2D_Reissner_Mindlin_plate_analytical(ncoor, nen)	154
body_gl_2D_Kirschhoff_Plate(nodxy, elenod, nen)	215
body_gl_2D_Plane(nodxy, elenod, nen, body)	254
body_gl_2D_Reissner_Mindlin_plate(nodxy, elenod, nen)	293
f_edges_2D_plane_analytical(ncoor, pressures_e, nen, edges_e)	339
f_edges_2D_plane_numerical(ncoor, pressures, nen, edges)	405
f_el_2D_plane_analytical(ncoor, pressures, nen, edges)	477
f_el_2D_plane_numerical(ncoor, pressures, nen, edges)	557
FlatNodePartVector(nv)	624
k_el_2D_Kirschhoff_plate_analytical(ncoor, mat, nen)	653
k_el_2D_Kirschhoff_plate_numerical(ncoor, mat, nen, nqp)	814
k_el_2D_plane_analytical(ncoor, mat, nen, type)	902
k_el_2D_plane_numerical(ncoor, mat, nqp, nen, type)	1000
k_el_2D_Reissner_Mindlin_plate_analytical(ncoor, mat, nen)	1102
k_el_2D_Reissner_Mindlin_plate_numerical(ncoor, mat, nqp, nen)	1187
k_gl_2D_Kirschhoff_Plate(nodxy, elenod, elemat, nen, type, numerical)	1265
k_gl_2D_plane(nodxy, elenod, elemat, nen, type, numerical)	1312
k_gl_2D_Reissner_Mindlin_plate(nodxy, elenod, elemat, nen, type, numerical)	1360
KirschhoffPlateElementsSolution(nodxy, elenod, elemat, nodtag, nodval, numerical)	1412
ModifiedMasterStiffness(nodtag, K)	1448
ModifiedNodeForces(nodtag, nodval, K, f)	1472
newsolver(Kmod, fmod)	1503
PlanarElementsSolution(nodxy, elenod, elemat, edges, pressures, body, nodtag, nodval, numerical)	1539
polynomial_maker(nen, xi)	1576
PrescDispDOFTags(nodtag)	1597
PrescDispDOFValues(nodtag, nodval)	1617
quadrature_1d(nqp)	1636
quadrature_2d(nqp)	1708
ReissnerMindlinPlateElementsSolution(nodxy, elenod, elemat, nodtag, nodval, numerical)	1845
shape_1d_nNodes_rod(nen, ncoor_temp, xi_val)	1887
shape_functions_2d(point, ncoor)	1936
shape_functions_2d_Kirschhoff_plate(point, ncoor)	2026
strain_el_2D_plane_numerical(ncoor, nqp, nen, displ, type)	2207
stress_el_2D_plane_numerical(mat, nqp, strain, type)	2266
symmetric_quadrature_2d(nqp)	2311
triangular_f_edges_2D_plane_analytical(ncoor, pressures_e, nen, edges_e)	2390
triangular_shape_functions_2d(point, ncoor)	2466

```
bf_el       : elemental body force matrix
*/
body_2D_plane_analytical(ncoor, nen, body) :=
block([current_point, coef, i, l, xi, eta],

  /* initialize the elemental stiffness matrix */
  b_force : zeromatrix(2*nen, 1),

  /* store the current quadrature point coordinates */
  current_point[1] : xi,
  current_point[2] : eta,

  /* calculate the shape functions and their derivatives at the current
     quadrature point*/
  if (nen = 4) then (
    [shape, jac] : shape_functions_2d(current_point, ncoor) )
  elseif (nen = 3) then (
    [shape, jac] : triangular_shape_functions_2d(current_point, ncoor)
  ),

  /* calculation of B-matrix */
  B_mat : zeromatrix(nen*2,2),
  for i : 1 thru nen do (
    B_mat[(i-1)*2+1,1] : shape[3,i],
    B_mat[(i-1)*2+2,2] : shape[3,i]
  ),

  /* calculation of thickness */
  current_point_x : 0,
  current_point_y : 0,
  for i : 1 thru nen do(
    current_point_x : current_point_x + ncoor[i][1]*shape[3,i],
    current_point_y : current_point_y + ncoor[i][2]*shape[3,i]
  ),
  th : thickness(current_point_x,current_point_y),

  if (nen = 4) then (
    bf_el : integrate(integrate(th*(B_mat.body)*jac,eta,-1,1),xi,-1,1) )
  elseif (nen = 3) then (
    bf_el : integrate(integrate(th*(B_mat.body)*jac,xi,0,1-eta),eta,0,1)
  ),

  /* return the elemental stiffness matrix */
  return(bf_el)
)$

/*
2D linear elasticity element
Classical Plate Elements
elemental force matrix
analytical integration

Inputs:
ncoor(2,4)  : nodal point coordinates
          ncoor[1,:] : x-direction
          ncoor[2,:] : y-direction

nen         : number of nodes per element

```

```
pressures  : pressures in the x- and y-directions

Output:

bf_el : elemental body force matrix

*/
body_el_2D_Kirschhoff_plate_analytical(ncoor, nen) :=
block([],

  /* initialize the elemental stiffness matrix */
  bf_el : zeromatrix(3*nen, 1),

  /* store the current quadrature point coordinates */
  current_point[1] : xi,
  current_point[2] : eta,

  /* calculate the shape functions and their derivatives at the current
     quadrature point*/
  if (nen = 4) then (
    /* calculate the shape functions and their derivatives at the current
     quadrature point*/
    [shape, shape2s, jac, shape_l] :
     shape_functions_2d_Kirschhoff_plate(current_point, ncoor) )
  elseif (nen = 3) then (
    /* to be added later */
    print("not programmed yet")
  ),

  /* calculation of B-matrix */
  N_mat : zeromatrix(nen*3,1),
  for i : 1 thru nen do (
    N_mat[(i-1)*3+1,1] : shape[3,1,i],
    N_mat[(i-1)*3+2,1] : shape[3,2,i],
    N_mat[(i-1)*3+3,1] : shape[3,3,i]
  ),

  /* calculation of thickness */
  current_point_x : 0,
  current_point_y : 0,
  for i : 1 thru nen do(
    current_point_x : current_point_x + ncoor[i][1]*shape_l[3,i],
    current_point_y : current_point_y + ncoor[i][2]*shape_l[3,i]
  ),
  qz : body_force(current_point_x,current_point_y),

  if (nen = 4) then (
    bf_el : integrate(integrate(ratsimp((N_mat*qz)*jac),eta,-1,1),xi,-1,1) )
  elseif (nen = 3) then (
    /* bf_el : integrate(integrate(th*(B_mat.body)*jac,xi,0,1-eta),eta,0,1) */
    print("not programmed yet")
  ),

  /* return the elemental stiffness matrix */
  return(ratsimp(bf_el))

)$

/*
```

```
2D Reissner Mindlin Plate Element
analytical integration

Inputs:
ncoor(2,4)  : nodal point coordinates
          ncoor[1,:] : x-direction
          ncoor[2,:] : y-direction

nen         : number of nodes per element

body        : body forces

Output:

f_el : elemental force matrix

*/
body_el_2D_Reissner_Mindlin_plate_analytical(ncoor, nen) :=
block([],

  /* initialize the elemental stiffness matrix */
  bf_el : zeromatrix(3*nen, 1),

  /* store the current quadrature point coordinates */
  current_point[1] : xi,
  current_point[2] : eta,

  /* calculate the shape functions and their derivatives at the current
     quadrature point*/
  if (nen = 4) then (
    /* calculate the shape functions and their derivatives at the current
     quadrature point*/
    [shape, jac] : shape_functions_2d(current_point, ncoor) )
  elseif (nen = 3) then (
    /* to be added later */
    print("not programmed yet")
  ),

  /* calculation of B-matrix */
  N_mat : zeromatrix(nen*3,3),
  for i : 1 thru nen do (
    N_mat[(i-1)*3+1,1] : shape[3,i],
    N_mat[(i-1)*3+2,2] : shape[3,i],
    N_mat[(i-1)*3+3,3] : shape[3,i]
  ),

  /* calculation of thickness */
  current_point_x : 0,
  current_point_y : 0,
  for i : 1 thru nen do(
    current_point_x : current_point_x + ncoor[i][1]*shape[3,i],
    current_point_y : current_point_y + ncoor[i][2]*shape[3,i]
  ),
  qz : q_z(current_point_x,current_point_y),
  mx : m_x(current_point_x,current_point_y),
  my : m_y(current_point_x,current_point_y),

  body : transpose(matrix([qz,mx,my])),
```

```
  bf_el : integrate(integrate(ratsimp((N_mat.body)*jac),eta,-1,1),xi,-1,1),

  /* return the elemental stiffness matrix */
  return(ratsimp(bf_el))

)$

/*
Global (Master) Matrix for Body Forces
2D Plate Problem

Inputs:
  nodxz   Nodal coordinates
  elenod  Element end nodes
  lint    Number of quadrature points
  nen     Number of element nodes

Output:
  fbody   Global matrix of body forces

*/
body_gl_2D_Kirschhoff_Plate(nodxy, elenod, nen) :=
block([numele,numnod,neldof,e,eftab,i,j,ii,jj,ncoor,Ke,K,K_els],
  numele : length(elenod),
  numnod : length(nodxy),
  fbody : zeromatrix(3*numnod, 1),
  for e:1 thru numele do (
    eftab : makelist(0,nen*3),
    ncoor : makelist(0,nen),
    for i:1 thru nen do(
      eftab[3*i-2] : 3*elenod[e][i]-2,
      eftab[3*i-1] : 3*elenod[e][i]-1,
      eftab[3*i] : 3*elenod[e][i],
      ncoor[i] : nodxy[elenod[e][i]]
      ),
    kill(fbody_el),
    fbody_el : body_el_2D_Kirschhoff_plate_analytical(ncoor, nen),
    neldof: length(fbody_el),
    i1 : 0,
    for i:1 thru neldof do(
      ii : eftab[i],
      fbody[ii,1] : fbody[ii,1]+fbody_el[i,1]
    )
  ),
  return(fbody)
)$

/*
Global (Master) Body Forces Matrix
2D Planar Problem

Inputs:
  nodxz   Nodal coordinates
  elenod  Element end nodes
  lint    Number of quadrature points
  nen     Number of element nodes
Output:
  K_els   Local stiffness matrix for individual elements
  K       Global (Master) Stiffness matrix
```

```
*/
body_gl_2D_Plane(nodxy, elenod, nen, body) :=
block([numele,numnod,neldof,e,eftab,i,j,ii,jj,ncoor,Ke,K,K_els],
  numele : length(elenod),
  numnod : length(nodxy),
  fbody : zeromatrix(2*numnod, 1),
  for e:1 thru numele do (
    eftab : makelist(0,nen*2),
    ncoor : makelist(0,nen),
    for i:1 thru nen do(
      eftab[2*i-1] : 2*elenod[e][i]-1,
      eftab[2*i] : 2*elenod[e][i],
      ncoor[i] : nodxy[elenod[e][i]]
      ),
    kill(fbody_el),
    fbody_el : body_2D_plane_analytical(ncoor, nen, body),
    neldof: length(fbody_el),
    i1 : 0,
    for i:1 thru neldof do(
      ii : eftab[i],
      fbody[ii,1] : fbody[ii,1]+fbody_el[i,1]
    )
  ),
  return(fbody)
)$

/*
Global (Master) Matrix for Body Forces
2D Reissner Mindlin Plate Problem

Inputs:
  nodxz   Nodal coordinates
  elenod  Element end nodes
  lint    Number of quadrature points
  nen     Number of element nodes

Output:
  fbody   Global matrix of body forces

*/
body_gl_2D_Reissner_Mindlin_plate(nodxy, elenod, nen) :=
block([numele,numnod,neldof,e,eftab,i,j,ii,jj,ncoor,Ke,K,K_els],
  numele : length(elenod),
  numnod : length(nodxy),
  fbody : zeromatrix(3*numnod, 1),
  for e:1 thru numele do (
    eftab : makelist(0,nen*3),
    ncoor : makelist(0,nen),
    for i:1 thru nen do(
      eftab[3*i-2] : 3*elenod[e][i]-2,
      eftab[3*i-1] : 3*elenod[e][i]-1,
      eftab[3*i] : 3*elenod[e][i],
      ncoor[i] : nodxy[elenod[e][i]]
      ),
    kill(fbody_el),
    fbody_el : body_el_2D_Reissner_Mindlin_plate_analytical(ncoor, nen),
    neldof: length(fbody_el),
    i1 : 0,
    for i:1 thru neldof do(
```

```
    ii : eftab[i],
    fbody[ii,1] : fbody[ii,1]+fbody_el[i,1]
  )
 ),
 return(fbody)
)$

/*
2D linear elasticity element
elemental force matrix
pressure based
analytical integration

Inputs:
ncoor(2,4) : nodal point coordinates
          ncoor[1,:] : x-direction
          ncoor[2,:] : y-direction

nen        : number of nodes per element

pressures  : pressures in the x- and y-directions

Output:

f_el : elemental force matrix

*/
f_edges_2D_plane_analytical(ncoor, pressures_e, nen, edges_e) :=
block([],

  /* initialize the elemental stiffness matrix */
  f_el : zeromatrix(2*nen, 1),

  /* loop over all edges */
  for e : 1 thru length(edges_e) do (
    /* loop over all nodal pairs of the edges */
    for n : 1 thru (length(edges[e])-1) do (
      Px : pressures_e[e][1],
      Py : pressures_e[e][2],
      n1 : edges_e[e][n],
      n2 : edges_e[e][n+1],
      ncoor_temp : [ncoor[n1], ncoor[n2]],

      /* calculate the shape functions and their derivatives at the quadrature
     point*/
      [shape_temp, ~, jac_temp] : shape_1d_nNodes_rod(2,ncoor_temp,xi),

      /* calcualtion of N_temp */

      /* define an array to store N matrices temporarily */
      N_temp : zeromatrix(nen*2,1),

      N_temp[n1*2 - 1,1] : shape_temp[1,1]*Px,
      N_temp[n1*2,1] : shape_temp[1,1]*Py,
      N_temp[n2*2 - 1,1] : shape_temp[1,2]*Px,
      N_temp[n2*2,1] : shape_temp[1,2]*Py,

      /* calculation of thickness on the current edge */
      current_point_x : 0,
```

```
        current_point_y : 0,
        for i : 1 thru 2 do(
          current_point_x : current_point_x + ncoor_temp[i][1]*shape_temp[1,i],
          current_point_y : current_point_y + ncoor_temp[i][2]*shape_temp[1,i]
        ),
        th : thickness(current_point_x,current_point_y),

        f_el : f_el + integrate(th*(N_temp)*(jac_temp),xi,-1,1)
      )
    ),
    /* return the elemental force matrix */
    return(f_el)
)$

/*
2D linear elasticity element
elemental force matrix
pressure based
numerical integration
(for linear elements)

Inputs:
ncoor(2,4) : nodal point coordinates
             ncoor[1,:] : x-direction
             ncoor[2,:] : y-direction

nen          : number of nodes per element

pressures    : pressures in the x- and y-directions

Output:

f_el : elemental force matrix

*/
f_edges_2D_plane_numerical(ncoor, pressures, nen, edges) :=
block([e],

    /* initialize the elemental stiffness matrix */
    f_el : zeromatrix(2*nen, 1),

    /* calculate the quadrature points */
    qps : quadrature_1d(2),

    /* loop over all edges */
    for e : 1 thru length(edges) do (
      /* loop over all nodal pairs of the edges */
      for n : 1 thru (length(edges[e])-1) do (
        kill(Px,Py,n1,n2,ncoor_temp),
        Px : pressures[e][1],
        Py : pressures[e][2],
        n1 : edges[e][n],
        n2 : edges[e][n+1],
        ncoor_temp : [ncoor[n1], ncoor[n2]],
        /* loop over quadrature points */
        for l : 1 thru 2 do (

          /* calculate the shape functions and their derivatives at the current
      quadrature point*/
```

```
        [shape_temp, ~, jac_temp] : shape_1d_nNodes_rod(2,ncoor_temp,qps[1,1]),

        /* calcualtion of N_temp */

        /* define an array to store N matrices temporarily */
        N_temp : zeromatrix(nen*2,1),

        N_temp[n1*2 - 1,1] : shape_temp[1,1]*Px,
        N_temp[n1*2,1] : shape_temp[1,1]*Py,
        N_temp[n2*2 - 1,1] : shape_temp[1,2]*Px,
        N_temp[n2*2,1] : shape_temp[1,2]*Py,

        /* calculation of thickness on the current edge */
        current_point_x : 0,
        current_point_y : 0,
        for i : 1 thru 2 do(
          current_point_x : current_point_x + ncoor_temp[i][1]*shape_temp[1,i],
          current_point_y : current_point_y + ncoor_temp[i][2]*shape_temp[1,i]
        ),
        th : thickness(current_point_x,current_point_y),

        f_el : f_el + (th*(N_temp)*jac_temp*qps[1,2])
      )
    )
  ),
  /* return the elemental force matrix */
  return(f_el)
)$

/*
2D linear elasticity element
elemental force matrix
pressure based
analytical integration

Inputs:
ncoor(2,4) : nodal point coordinates
          ncoor[1,:] : x-direction
          ncoor[2,:] : y-direction

nen        : number of nodes per element

pressures  : pressures in the x- and y-directions

Output:

f_el : elemental force matrix

*/
f_el_2D_plane_analytical(ncoor, pressures, nen, edges) :=
block([],

  /* initialize the elemental stiffness matrix */
  f_el : zeromatrix(2*nen, 1),

  /* store the current quadrature point coordinates */
  current_point[1] : xi,
  current_point[2] : eta,

```

```
  /* calculate the shape functions and their derivatives at the current
     quadrature point */
  [shape, ~] : shape_functions_2d(current_point, ncoor),

  /* define edges */
  edge[1] : [xi, eta=-1],  /* 1-2 */
  edge[2] : [eta, xi=+1],  /* 2-3 */
  edge[3] : [xi, eta=+1],  /* 3-4 */
  edge[4] : [eta, xi=-1],  /* 4-1 */

  /* define an array to store N matrices temporarily */
  N_temp : zeromatrix(nen*2,1),

  /* loop over all edges */
  for e : 1 thru 4 do (
    Px : pressures[e][1],
    Py : pressures[e][2],

    /* calculate the shape functions and their derivatives at the current
     quadrature point*/
    for i : 1 thru nen do (
      shape_temp[3,i] : at(shape[3,i],edge[e][2])
    ),

    /* calculate the jacobian for the current edge */
    n1 : edges[e][1],
    n2 : edges[e][2],
    jac_temp : sqrt( (ncoor[n2][2]-ncoor[n1][2])^2 +
     (ncoor[n2][1]-ncoor[n1][1])^2 ),

    /* calcualtion of N_temp */
    for i : 1 thru nen do (
      N_temp[i*2 - 1,1] : shape_temp[3,i]*Px,
      N_temp[i*2,1] : shape_temp[3,i]*Py
    ),

    /* calculation of thickness on the current edge */
    current_point_x : 0,
    current_point_y : 0,
    for i : 1 thru nen do(
      current_point_x : current_point_x + ncoor[i][1]*shape_temp[3,i],
      current_point_y : current_point_y + ncoor[i][2]*shape_temp[3,i]
    ),
    th : thickness(current_point_x,current_point_y),

    f_el : f_el + integrate(th*(N_temp)*(jac_temp/2),edge[e][1],-1,1)

  ),

  /* return the elemental force matrix */
  return(f_el)
)$

/*
2D linear elasticity element
elemental force matrix
pressure based
numerical integration

```

```
Inputs:
ncoor(2,4) : nodal point coordinates
           ncoor[1,:] : x-direction
           ncoor[2,:] : y-direction

nen        : number of nodes per element

pressures  : pressures in the x- and y-directions

Output:

f_el : elemental force matrix

*/
f_el_2D_plane_numerical(ncoor, pressures, nen, edges) :=
block([],

  /* initialize the elemental stiffness matrix */
  f_el : zeromatrix(2*nen, 1),

  /* calculate the quadrature points */
  qps : quadrature_1d(2),

  /* define quadrature points on the edges */
  edge[1] : [[qps[1,1], -1],[qps[2,1], -1]],  /* 1-2 */
  edge[2] : [[+1, qps[1,1]],[+1, qps[2,1]]],  /* 2-3 */
  edge[3] : [[qps[2,1], +1],[qps[1,1], +1]],  /* 3-4 */
  edge[4] : [[-1, qps[2,1]],[-1, qps[1,1]]],  /* 4-1 */

  /* define an array to store N matrices temporarily */
  N_temp : zeromatrix(nen*2,1),

  /* loop over all edges */
  for e : 1 thru 4 do (
    Px : pressures[e][1],
    Py : pressures[e][2],

    for l : 1 thru 2 do (
      /* store the current quadrature point coordinates */
      current_point[1] : edge[e][l][1],
      current_point[2] : edge[e][l][2],

      /* calculate the shape functions and their derivatives at the current quadrature point*/
      kill(shape_temp,jac_temp),
      [shape_temp, ~] : shape_functions_2d(current_point, ncoor),

      /* calculate the jacobian for the current edge */
      n1 : edges[e][1],
      n2 : edges[e][2],
      jac_temp : sqrt( (ncoor[n2][2]-ncoor[n1][2])^2 + (ncoor[n2][1]-ncoor[n1][1])^2 ),

      /* calcualtion of N_temp */
      for i : 1 thru nen do (
        N_temp[i*2 - 1,1] : shape_temp[3,i]*Px,
        N_temp[i*2,1] : shape_temp[3,i]*Py
      ),

```

```
      /* calculation of thickness on the current edge */
      current_point_x : 0,
      current_point_y : 0,
      for i : 1 thru nen do(
        current_point_x : current_point_x + ncoor[i][1]*shape_temp[3,i],
        current_point_y : current_point_y + ncoor[i][2]*shape_temp[3,i]
      ),
      th : thickness(current_point_x,current_point_y),

      f_el : f_el + (th*(N_temp)*(jac_temp/2)*qps[1,2])
    )
  ),

  /* return the elemental force matrix */
  return(f_el)
)$

/*
Flattening an array
Inputs:
  nv  input array
Output:
  nv  flattened array
*/
FlatNodePartVector(nv) := flatten(nv)$

/*
2D linear elasticity element
Classical Plate Elements
elemental stiffness matrix
analytical integration

Inputs:
ncoor(2,4) : nodal point coordinates
          ncoor[1,:] : x-direction
          ncoor[2,:] : y-direction

nen       : number of nodes per element

type      : type of the element
          "plane stress" or "plane strain"

mat       : material properies
          mat[1] : Young's modulus
          mat[2] : Poison's ratio

Output:
shp(3,4)  : shape functions array
           shp(1,:) : derivatives in the x-direction
           shp(2,:) : derivatives in the y-direction
           shp(3,:) : shape functions

*/
k_el_2D_Kirschhoff_plate_analytical(ncoor, mat, nen) :=
block([current_point, coef, i, l, xi, eta,K_el],
  assume(xi>=-1,xi<=1,eta>=-1,eta<1),

  /* initialize the elemental stiffness matrix */
  K_el : zeromatrix(3*nen, 3*nen),
```

```
/* store the current quadrature point coordinates */
current_point[1] : xi,
current_point[2] : eta,

/* calculate the shape functions and their derivatives at the current
   quadrature point*/
if (nen = 4) then (
  [shape, shape2s, jac, shape_1] :
  shape_functions_2d_Kirschhoff_plate(current_point, ncoor) )
elseif (nen = 3) then (
  L12 : sqrt((ncoor[2][1]-ncoor[1][1])^2+(ncoor[2][2]-ncoor[1][2])^2),

  c : (ncoor[2][1]-ncoor[1][1])/L12,
  s : (ncoor[2][2]-ncoor[1][2])/L12,

  T_mat : matrix(
    [1,0,0,0,0,0,0,0,0],
    [0,c,-s,0,0,0,0,0,0],
    [0,s,c,0,0,0,0,0,0],
    [0,0,0,1,0,0,0,0,0],
    [0,0,0,0,c,-s,0,0,0],
    [0,0,0,0,s,c,0,0,0],
    [0,0,0,0,0,0,1,0,0],
    [0,0,0,0,0,0,0,c,-s],
    [0,0,0,0,0,0,0,s,c]
  ),

  TX_mat : matrix(
    [c,-s,0,0,0,0],
    [s, c,0,0,0,0],
    [0, 0,c,-s,0,0],
    [0, 0,s,c,0,0],
    [0, 0,0,0,c,-s],
    [0, 0,0,0,s,c]
  ),

  X_mat : matrix(
  [ncoor[1][1],ncoor[1][2],
  ncoor[2][1],ncoor[2][2],
  ncoor[3][1],ncoor[3][2]]),

  x_mat : TX_mat . X_mat,

  x1 : x_mat[1][1],
  y1 : x_mat[2][1],
  x2 : x_mat[3][1],
  y2 : x_mat[4][1],
  x3 : x_mat[5][1],
  y3 : x_mat[6][1],

  [shape, shape2s, jac, shape_1] :
   triangular_shape_functions_2d_Kirschhoff_plate_1(current_point, ncoor)
),

/* calculation of B-matrix */
B_mat : zeromatrix(nen*3,3),
ic : 0,
for i : 1 thru nen do (
```

```
      for j : 1 thru 3 do (
          ic : ic + 1,
          B_mat[ic,1] : ratsimp(factor(shape2s[1,j,i])),
          B_mat[ic,2] : ratsimp(factor(shape2s[2,j,i])),
          B_mat[ic,3] : ratsimp(factor(2*shape2s[3,j,i]))
      )
    ),
    /* calculation of BT_matrix */
    BT_mat : transpose(B_mat),

    /* calculation of thickness */
    current_point_x : 0,
    current_point_y : 0,
    for i : 1 thru nen do(
      current_point_x : current_point_x + ncoor[i][1]*shape_1[3,i],
      current_point_y : current_point_y + ncoor[i][2]*shape_1[3,i]
    ),
    th : thickness(current_point_x,current_point_y),

    /* calculation of the elasticity matrix */
    C_mat : zeromatrix(3,3),
    coef : mat[1]*(th^3)/(12*(1-mat[2]^2)),
    C_mat[1,1] : 1,
    C_mat[1,2] : mat[2],
    C_mat[2,1] : mat[2],
    C_mat[2,2] : 1,
    C_mat[3,3] : (1-mat[2])/2,
    C_mat : coef*C_mat,
    if (nen = 4) then (
      /*K_el : integrate(integrate(factor(factor(ratsimp(B_mat)).
      (C_mat.factor(ratsimp(BT_mat))))*jac,eta,-1,1),xi,-1,1) */
      K_el : realpart(integrate(factor(B_mat.(C_mat.BT_mat)*jac),eta,-1,1)),
      K_el : realpart(integrate(factor(K_el),xi,-1,1))
    ) elseif (nen = 3) then (

    A : (1/2)*(ncoor[1][1]*ncoor[2][2]-ncoor[2][1]*ncoor[1][2]+ncoor[2][1]
      *ncoor[3][2]-ncoor[3][1]*ncoor[2][2]+ncoor[3][1]*ncoor[1][2]
      -ncoor[1][1]*ncoor[3][1]),

    Xc : (ncoor[1][1]+ncoor[2][1]+ncoor[3][1])/3,
    Yc : (ncoor[1][2]+ncoor[2][2]+ncoor[3][2])/3,

    sum_x2 : (ncoor[3][1]^2 + (ncoor[2][1]+ncoor[1][1])*ncoor[3][1] +
       ncoor[2][1]^2 + ncoor[1][1]*ncoor[2][1] + ncoor[1][1]^2) / (
       2*ncoor[3][1]+2*ncoor[2][1]+2*ncoor[1][1] ),

    sum_y2 : (ncoor[3][2]^2 + (ncoor[2][2]+ncoor[1][2])*ncoor[3][2] +
       ncoor[2][2]^2 + ncoor[1][2]*ncoor[2][2] + ncoor[1][2]^2) / (
       2*ncoor[3][2]+2*ncoor[2][2]+2*ncoor[1][2] ),

    sum_xy : (
    (6*ncoor[3][1]+3*ncoor[2][1]+3*ncoor[1][1])*ncoor[3][2] +
    (3*ncoor[3][1]+6*ncoor[2][1]+3*ncoor[1][1])*ncoor[2][2] +
    (3*ncoor[3][1]+3*ncoor[2][1]+6*ncoor[1][1])*ncoor[1][2]   )
    / (
    (4*ncoor[3][1]+4*ncoor[2][1]+4*ncoor[1][1])*
    (ncoor[3][2]+ncoor[2][2]+ncoor[1][2])   ),

    int_1 : A,
```

```
  int_2 : A*Xc,
  int_3 : A*Yc,
  int_4 : A*Xc + sum_x2,
  int_5 : A*Yc + sum_y2,
  int_6 : Xc*Yc + sum_xy,

  K_temp : ratsubst(int_6,x*y,ratsimp(factor(B_mat.(C_mat.BT_mat)))),
  K_temp : subst([y^2=int_5,x^2=int_4,y=int_3,x=int_2],K_temp),

  K_el : ratsimp(factor(transpose(T_mat).(ratsimp(K_temp).T_mat)))

  ),

  /* return the elemental stiffness matrix */
  return(K_el)
)$

/*
2D linear elasticity element
Classical Plate Elements
elemental stiffness matrix
numerical integration

Inputs:
ncoor(2,4) : nodal point coordinates
           ncoor[1,:] : x-direction
           ncoor[2,:] : y-direction

nen      : number of nodes per element

type     : type of the element
           "plane stress" or "plane strain"

mat      : material properies
           mat[1] : Young's modulus
           mat[2] : Poison's ratio

Output:
shp(3,4) : shape functions array
            shp(1,:) : derivatives in the x-direction
            shp(2,:) : derivatives in the y-direction
            shp(3,:) : shape functions

*/
k_el_2D_Kirschhoff_plate_numerical(ncoor, mat, nen, nqp) :=
block([current_point, coef, i, l, xi, eta, K_el],

  /* initialize the elemental stiffness matrix */
  K_el : zeromatrix(3*nen, 3*nen),

  /* calculate the quadrature points */
  qps : quadrature_2d(nqp),

  /* loop over the quadrature points */
  for l : 1 thru nqp do(

    /* store the current quadrature point coordinates */
     current_point[1] : qps[1,1],
     current_point[2] : qps[1,2],
```

```
    /* calculate the shape functions and their derivatives at the current
     quadrature point*/
    [shape, shape2s, jac, shape_l] :
     shape_functions_2d_Kirschhoff_plate(current_point, ncoor),

    /* calculation of B-matrix */
    B_mat : zeromatrix(nen*3,3),
    ic : 0,
    for i : 1 thru nen do (
      for j : 1 thru 3 do (
        ic : ic + 1,
        B_mat[ic,1] : shape2s[1,j,i],
        B_mat[ic,2] : shape2s[2,j,i],
        B_mat[ic,3] : 2*shape2s[3,j,i]
      )
    ),

    /* calculation of BT_matrix */
    BT_mat : transpose(B_mat),

    /* calculation of thickness */
    current_point_x : 0,
    current_point_y : 0,
    for i : 1 thru nen do(
      current_point_x : current_point_x + ncoor[i][1]*shape_l[3,i],
      current_point_y : current_point_y + ncoor[i][2]*shape_l[3,i]
    ),
    th : thickness(current_point_x,current_point_y),

    /* calculation of the elasticity matrix */
    C_mat : zeromatrix(3,3),
    coef : mat[1]*(th^3)/(12*(1-mat[2]^2)),
    C_mat[1,1] : 1,
    C_mat[1,2] : mat[2],
    C_mat[2,1] : mat[2],
    C_mat[2,2] : 1,
    C_mat[3,3] : (1-mat[2])/2,
    C_mat : coef*C_mat,

    K_el : K_el + (B_mat.(C_mat.BT_mat))*jac*qps[1,3]

  ),

  /* return the elemental stiffness matrix */
  return(K_el)
)$

/*
2D linear elasticity element
plane stres and plane strain
elemental stiffness matrix
analytical integration

Inputs:
ncoor(2,4) : nodal point coordinates
          ncoor[1,:] : x-direction
          ncoor[2,:] : y-direction

```

```
nen      : number of nodes per element

type     : type of the element
           "plane stress" or "plane strain"

mat      : material properies
           mat[1] : Young's modulus
           mat[2] : Poison's ratio

Output:
shp(3,4) : shape functions array
            shp(1,:) : derivatives in the x-direction
            shp(2,:) : derivatives in the y-direction
            shp(3,:) : shape functions

*/
k_el_2D_plane_analytical(ncoor, mat, nen, type) :=
block([current_point, coef, i, l, xi, eta],

  /* initialize the elemental stiffness matrix */
  K_el : zeromatrix(2*nen, 2*nen),

  /* store the current quadrature point coordinates */
  current_point[1] : xi,
  current_point[2] : eta,

  /* calculate the shape functions and their derivatives at the current
     quadrature point*/
  if (nen = 4) then (
    [shape, jac] : shape_functions_2d(current_point, ncoor) )
  elseif (nen = 3) then (
    [shape, jac] : triangular_shape_functions_2d(current_point, ncoor)
  ),

  /* calculation of B-matrix */
  B_mat : zeromatrix(nen*2,3),
  for i : 1 thru nen do (
    B_mat[(i-1)*2+1,1] : shape[1,i],
    B_mat[(i-1)*2+1,3] : shape[2,i],
    B_mat[(i-1)*2+2,2] : shape[2,i],
    B_mat[(i-1)*2+2,3] : shape[1,i]
  ),

  /* calculation of BT_matrix */
  BT_mat : transpose(B_mat),

  /* calculation of the elasticity matrix */
  if (type = "plane stress") then (
    C_mat : zeromatrix(3,3),
    coef : mat[1]/(1-mat[2]^2),
    C_mat[1,1] : 1,
    C_mat[1,2] : mat[2],
    C_mat[2,1] : mat[2],
    C_mat[2,2] : 1,
    C_mat[3,3] : (1-mat[2])/2,
    C_mat : coef*C_mat
  ) elseif (type = "plane strain") then (
    C_mat : zeromatrix(3,3),
      coef : mat[1]/((1+mat[2])*(1-2*mat[2])),
```

```
        C_mat[1,1] : 1-mat[2],
        C_mat[1,2] : mat[2],
        C_mat[2,1] : mat[2],
        C_mat[2,2] : 1-mat[2],
        C_mat[3,3] : (1-2*mat[2])/2,
        C_mat : coef*C_mat
    ),

    /* calculation of thickness */
    current_point_x : 0,
    current_point_y : 0,
    for i : 1 thru nen do(
      current_point_x : current_point_x + ncoor[i][1]*shape[3,i],
      current_point_y : current_point_y + ncoor[i][2]*shape[3,i]
    ),
    th : thickness(current_point_x,current_point_y),

    if (nen = 4) then (
      K_el :
       integrate(integrate(th*(B_mat.(C_mat.BT_mat))*jac,eta,-1,1),xi,-1,1) )
    elseif (nen = 3) then (
      K_el :
       integrate(integrate(th*(B_mat.(C_mat.BT_mat))*jac,eta,0,1-xi),xi,0,1)
    ),

    /* return the elemental stiffness matrix */
    return(K_el)
)$

/*
2D linear elasticity element
plane stres and plane strain
elemental stiffness matrix
numerical integration

Inputs:
ncoor[4][2] : nodal point coordinates
              ncoor[:][1] : x-direction
              ncoor[:][2] : y-direction

nqp        : number of quadrature points per direction

nen         : number of nodes per element

type        : type of the element
              "plane stress" or "plane strain"

mat         : material properies
              mat[1] : Young's modulus
              mat[2] : Poison's ratio

Output:
shp(3,4)    : shape functions array
              shp(1,:) : derivatives in the x-direction
              shp(2,:) : derivatives in the y-direction
              shp(3,:) : shape functions

*/
k_el_2D_plane_numerical(ncoor, mat, nqp, nen, type) :=
```

```
block([current_point, coef, i, l],

  /* initialize the elemental stiffness matrix */
  K_el : zeromatrix(2*nen, 2*nen),

  if (nen = 4) then (
    /* calculate the quadrature points */
    qps : quadrature_2d(nqp))
  elseif (nen = 3) then (
    qps : bfloat(symmetric_quadrature_2d(nqp))
  ),

  /* loop over the quadrature points */
  for l : 1 thru nqp do(

    /* store the current quadrature point coordinates */
    current_point[1] : qps[l,1],
    current_point[2] : qps[l,2],

    /* calculate the shape functions and their derivatives at the current
     quadrature point*/
    if (nen = 4) then (
      [shape, jac] : shape_functions_2d(current_point, ncoor) )
    elseif (nen = 3) then (
      [shape, jac] : triangular_shape_functions_2d(current_point, ncoor)
    ),

    /* calculation of B-matrix */
    B_mat : zeromatrix(nen*2,3),
    for i : 1 thru nen do (
      B_mat[(i-1)*2+1,1] : shape[1,i],
      B_mat[(i-1)*2+1,3] : shape[2,i],
      B_mat[(i-1)*2+2,2] : shape[2,i],
      B_mat[(i-1)*2+2,3] : shape[1,i]
    ),

    /* calculation of BT_matrix */
    BT_mat : transpose(B_mat),

    /* calculation of the elasticity matrix */
    C_mat : zeromatrix(3,3),
    if (type = "plane stress") then (
      coef : mat[1]/(1-mat[2]^2),
      C_mat[1,1] : 1.0,
      C_mat[1,2] : mat[2],
      C_mat[2,1] : mat[2],
      C_mat[2,2] : 1.0,
      C_mat[3,3] : (1-mat[2])/2.0,
      C_mat : coef*C_mat
    ) elseif (type = "plane strain") then (
      coef : mat[1]/((1+mat[2])*(1-2*mat[2])),
      C_mat[1,1] : 1-mat[2],
      C_mat[1,2] : mat[2],
      C_mat[2,1] : mat[2],
      C_mat[2,2] : 1-mat[2],
      C_mat[3,3] : (1-2*mat[2])/2.0,
      C_mat : coef*C_mat
    ),

```

```
      /* calculation of thickness */
      current_point_x : 0,
      current_point_y : 0,
      for i : 1 thru nen do(
      current_point_x : current_point_x + ncoor[i][1]*shape[3,i],
      current_point_y : current_point_y + ncoor[i][2]*shape[3,i]
      ),
      th : thickness(current_point_x,current_point_y),
      K_el : K_el + th*(B_mat.(C_mat.BT_mat))*jac*qps[1,3]

    ),

    /* return the elemental stiffness matrix */
    return(K_el)
)$

/*
2D linear elasticity element
plane stres and plane strain
elemental stiffness matrix
analytical integration

Inputs:
ncoor(2,4) : nodal point coordinates
          ncoor[1,:] : x-direction
          ncoor[2,:] : y-direction

nen      : number of nodes per element

type     : type of the element
          "plane stress" or "plane strain"

mat      : material properies
          mat[1] : Young's modulus
          mat[2] : Poison's ratio

Output:
shp(3,4) : shape functions array
           shp(1,:) : derivatives in the x-direction
           shp(2,:) : derivatives in the y-direction
           shp(3,:) : shape functions

*/
k_el_2D_Reissner_Mindlin_plate_analytical(ncoor, mat, nen) :=
block([current_point, coef, i, l, xi, eta],

  /* initialize the elemental stiffness matrix */
  K_el : zeromatrix(3*nen, 3*nen),

  /* store the current quadrature point coordinates */
  current_point[1] : xi,
  current_point[2] : eta,

  /* calculate the shape functions and their derivatives at the current
     quadrature point*/
  [shape, jac] : shape_functions_2d(current_point, ncoor),

  /* calculation of B-matrix */
  B_mat : zeromatrix(nen*3,5),
```

```
  for i : 1 thru nen do (
    B_mat[(i-1)*3+1,4] : shape[1,i],
    B_mat[(i-1)*3+1,5] : shape[2,i],
    B_mat[(i-1)*3+2,2] : -1*shape[2,i],
    B_mat[(i-1)*3+2,3] : -1*shape[1,i],
    B_mat[(i-1)*3+2,5] : -1*shape[3,i],
    B_mat[(i-1)*3+3,1] : shape[1,i],
    B_mat[(i-1)*3+3,3] : shape[2,i],
    B_mat[(i-1)*3+3,4] : shape[3,i]
  ),

  /* calculation of BT_matrix */
  BT_mat : transpose(B_mat),

  /* calculation of thickness */
  current_point_x : 0,
  current_point_y : 0,
  for i : 1 thru nen do(
    current_point_x : current_point_x + ncoor[i][1]*shape[3,i],
    current_point_y : current_point_y + ncoor[i][2]*shape[3,i]
  ),
  th : thickness(current_point_x,current_point_y),

  /* calculation of the elasticity matrix */
  D_mat : zeromatrix(5,5),

  D_mat[1,1] : D_b,
  D_mat[1,2] : D_b*mat[2],
  D_mat[2,1] : D_b*mat[2],
  D_mat[2,2] : D_b,
  D_mat[3,3] : D_b*((1-mat[2])/2),

  D_mat[4,4] : D_s,
  D_mat[5,5] : D_s,


  K_el : integrate(integrate((B_mat.(D_mat.BT_mat))*jac,eta,-1,1),xi,-1,1),

  /* return the elemental stiffness matrix */
  return(K_el)
)$

/*
2D Reissner Mindlin Plate Element
elemental stiffness matrix
numerical integration

Inputs:
ncoor[4][2] : nodal point coordinates
              ncoor[:][1] : x-direction
              ncoor[:][2] : y-direction

nqp         : number of quadrature points per direction

nen          : number of nodes per element

type         : type of the element
               "plane stress" or "plane strain"

```

```
mat          : material properies
               mat[1] : Young's modulus
               mat[2] : Poison's ratio

Output:
shp(3,4)     : shape functions array
               shp(1,:) : derivatives in the x-direction
               shp(2,:) : derivatives in the y-direction
               shp(3,:) : shape functions

*/
k_el_2D_Reissner_Mindlin_plate_numerical(ncoor, mat, nqp, nen) :=
block([current_point, coef, i, l],

  /* initialize the elemental stiffness matrix */
  K_el : zeromatrix(3*nen, 3*nen),

  /* calculate the quadrature points */
  qps : (quadrature_2d(nqp)),

  /* loop over the quadrature points */
  for l : 1 thru nqp do(

    /* store the current quadrature point coordinates */
    current_point[1] : qps[l,1],
    current_point[2] : qps[l,2],

    /* calculate the shape functions and their derivatives at the current
     quadrature point*/
    [shape, jac] : shape_functions_2d(current_point, ncoor),

    /* calculation of B-matrix */
    B_mat : zeromatrix(nen*3,5),
    for i : 1 thru nen do (
      B_mat[(i-1)*3+1,4] : shape[1,i],
      B_mat[(i-1)*3+1,5] : shape[2,i],
      B_mat[(i-1)*3+2,2] : -1*shape[2,i],
      B_mat[(i-1)*3+2,3] : -1*shape[1,i],
      B_mat[(i-1)*3+2,5] : -1*shape[3,i],
      B_mat[(i-1)*3+3,1] : shape[1,i],
      B_mat[(i-1)*3+3,3] : shape[2,i],
      B_mat[(i-1)*3+3,4] : shape[3,i]
    ),

    /* calculation of BT_matrix */
    BT_mat : transpose(B_mat),

    /* calculation of thickness */
    current_point_x : 0,
    current_point_y : 0,
    for i : 1 thru nen do(
    current_point_x : current_point_x + ncoor[i][1]*shape[3,i],
    current_point_y : current_point_y + ncoor[i][2]*shape[3,i]
    ),
    th : thickness(current_point_x,current_point_y),

    /* calculation of the elasticity matrix */
    D_mat : zeromatrix(5,5),

```

```
    D_mat[1,1] : D_b,
    D_mat[1,2] : D_b*mat[2],
    D_mat[2,1] : D_b*mat[2],
    D_mat[2,2] : D_b,
    D_mat[3,3] : D_b*((1-mat[2])/2),

    D_mat[4,4] : D_s,
    D_mat[5,5] : D_s,


    K_el : K_el + (B_mat.(D_mat.BT_mat))*jac*qps[1,3]

  ),

  /* return the elemental stiffness matrix */
  return(K_el)
)$

/*
Global (Master) Stiffness Matrix
2D Plate Problem

Inputs:
  nodxz   Nodal coordinates
  elenod  Element end nodes
  lint    Number of quadrature points
  nen     Number of element nodes
Output:
  K_els   Local stiffness matrix for individual elements
  K       Global (Master) Stiffness matrix
*/
k_gl_2D_Kirschhoff_Plate(nodxy, elenod, elemat, nen, type, numerical) :=
block([numele,numnod,neldof,e,eftab,i,j,ii,jj,ncoor,Ke,K,K_els],
  numele : length(elenod),
  numnod : length(nodxy),
  K : zeromatrix(3*numnod, 3*numnod),
  for e:1 thru numele do (
    eftab : makelist(0,nen*3),
    ncoor : makelist(0,nen),
    for i:1 thru nen do(
      eftab[3*i-2] : 3*elenod[e][i]-2,
      eftab[3*i-1] : 3*elenod[e][i]-1,
      eftab[3*i] : 3*elenod[e][i],
      ncoor[i] : nodxy[elenod[e][i]],
      mat : elemat[e]
    ),
    if (numerical) then (
      Ke : k_el_2D_Kirschhoff_plate_numerical(ncoor, mat, nen, nqp) )
    else (
      Ke : k_el_2D_Kirschhoff_plate_analytical(ncoor, mat, nen)
    ),
    K_els[e] :  Ke,
    neldof: length(Ke),
    i1 : 0,
    for i:1 thru neldof do(
      ii : eftab[i],
      for j:1 thru neldof do(
        jj : eftab[j],
        K[ii,jj] : K[ii,jj]+Ke[i,j]
```

```
        )
      )
    ),
    return([K_els,K])
  )$

/*
Global (Master) Stiffness Matrix
2D Planar Problem

Inputs:
  nodxz    Nodal coordinates
  elenod   Element end nodes
  lint     Number of quadrature points
  nen      Number of element nodes
Output:
  K_els    Local stiffness matrix for individual elements
  K        Global (Master) Stiffness matrix
*/
k_gl_2D_plane(nodxy, elenod, elemat, nen, type, numerical) :=
block([numele,numnod,neldof,e,eftab,i,j,ii,jj,ncoor,Ke,K,K_els],
  numele : length(elenod),
  numnod : length(nodxy),
  K : zeromatrix(2*numnod, 2*numnod),
  for e:1 thru numele do (
    eftab : makelist(0,nen*2),
    ncoor : makelist(0,nen),
    for i:1 thru nen do(
      eftab[2*i-1] : 2*elenod[e][i]-1,
      eftab[2*i] : 2*elenod[e][i],
      ncoor[i] : nodxy[elenod[e][i]],
      mat : elemat[e]
    ),
    if (numerical) then (
      Ke : k_el_2D_plane_numerical(ncoor, mat, nqp, nen, type) )
    else (
      Ke : k_el_2D_plane_analytical(ncoor, mat, nen, type)
    ),
    K_els[e] :  Ke,
    neldof: length(Ke),
    i1 : 0,
    for i:1 thru neldof do(
      ii : eftab[i],
      for j:1 thru neldof do(
        jj : eftab[j],
        K[ii,jj] : K[ii,jj]+Ke[i,j]
      )
    )
  ),
  return([K_els,K])
)$

/*
Global (Master) Stiffness Matrix
2D Reissner Mindlin Plate Problem

Inputs:
  nodxz    Nodal coordinates
  elenod   Element end nodes
```

```
  lint    Number of quadrature points
  nen     Number of element nodes

Output:
  K_els   Local stiffness matrix for individual elements
  K       Global (Master) Stiffness matrix

*/
k_gl_2D_Reissner_Mindlin_plate(nodxy, elenod, elemat, nen, type, numerical) :=
block([numele,numnod,neldof,e,eftab,i,j,ii,jj,ncoor,Ke,K,K_els],
  numele : length(elenod),
  numnod : length(nodxy),
  K : zeromatrix(3*numnod, 3*numnod),
  for e:1 thru numele do (
    eftab : makelist(0,nen*3),
    ncoor : makelist(0,nen),
    for i:1 thru nen do(
      eftab[3*i-2] : 3*elenod[e][i]-2,
      eftab[3*i-1] : 3*elenod[e][i]-1,
      eftab[3*i] : 3*elenod[e][i],
      ncoor[i] : nodxy[elenod[e][i]],
      mat : elemat[e]
      ),
      if (numerical) then (
        Ke : k_el_2D_Reissner_Mindlin_plate_numerical(ncoor, mat, nqp, nen) )
      else (
        Ke : k_el_2D_Reissner_Mindlin_plate_analytical(ncoor, mat, nen)
      ),
    K_els[e] :  Ke,
    neldof: length(Ke),
    i1 : 0,
    for i:1 thru neldof do(
      ii : eftab[i],
      for j:1 thru neldof do(
        jj : eftab[j],
        K[ii,jj] : K[ii,jj]+Ke[i,j]
      )
    )
  ),
  return([K_els,K])
)$

/*
2D Plate Problem - Solution Driver

Inputs:
  nodxy   Nodal coordinates
  elenod  Element end nodes
  elemat  Element material properties
  elefab  Element fabrication properties
  nodtag  Nodal freedom tags list
  nodval  Nodal freedom values list

Outputs:
  noddis  Nodal displacements
  nodfor  Nodal forces
  elefor  Element forces
  elesig  Element stresses

```

```
*/
KirschhoffPlateElementsSolution(nodxy, elenod, elemat, nodtag, nodval,
     numerical) :=
block([K,Kmod,f,fmod,u,noddis,nodfor,elefor,elesig,eftab,Ke,i,c],
  /* calculate the global stiffness matrix */
  [K_els, K] : k_gl_2D_Kirschhoff_Plate(nodxy, elenod, elemat, nen, type,
     numerical),
  /* calculate the modified global stiffness matrix */
  Kmod : ModifiedMasterStiffness(nodtag,K),
  /* calculate the global force matrix */
  f : transpose(FlatNodePartVector(nodval)),
  f : f + ratsimp(body_gl_2D_Kirschhoff_Plate(nodxy, elenod, nen)),

  /* calculate the modified global force matrix */
  fmod : ModifiedNodeForces(nodtag,nodval,K,f),


  /* solve the system of equation for displacements */
  u : invert(Kmod).fmod,

  /* transform u to noddis (distinguished nodal displacements) */
  noddis : zeromatrix(length(u)/3,3),
  c : 0,
  for i: 1 thru (length(u)/3) do(
    c : c + 1,
    noddis[i] : args(append(u[(c-1)*3+1],u[(c-1)*3+2],u[(c-1)*3+3]))
  ),
  return([float(noddis),Kmod,fmod])
)$


/*
Modified Global (Master) Stiffness Matrix
Inputs:
  nodtag  Nodal freedom tags list
  K       Global (master) stiffness matrix
Output:
  Kmod    Modified global (master) stiffness matrix
*/
ModifiedMasterStiffness(nodtag, K) :=
block([i,j,k,n:length(K),pdof,np,Kmod:copymatrix(K)],
  pdof : PrescDispDOFTags(nodtag),
  np : length(pdof),
  for k : 1 thru np do (
    i : pdof[k],
    for j : 1 thru n do (
      Kmod [i,j] : Kmod [j,i] : 0
    ),
    Kmod[i,i] : 1
  ),
  return(Kmod)
)$

/*
Modified Nodal Forces Matrix
Inputs:
  nodtag  Nodal freedom tags list
  nodval  Nodal freedom values list
  K       Global (master) stiffness matrix
```

```
  f       Global (master) force matrix
Output:
  fmod    Modified global (master) force matrix
*/
ModifiedNodeForces(nodtag, nodval, K, f) :=
block([i,j,k,n:length(K),pdof,pval,np,d,c,fmod:copymatrix(f)],
  pdof : PrescDispDOFTags(nodtag),
  np : length(pdof),
  pval : PrescDispDOFValues(nodtag, nodval),
  c : makelist(1,n),
  for k : 1 thru np do (
    i : pdof[k],
    c[i] : 0
  ),
  for k : 1 thru np do (
    i : pdof[k],
    d : pval[k],
    fmod[i,1] : d,
    if (d#0) then (
    for j : 1 thru n do (
      fmod[j,1] : fmod[j,1] - K[i,j]*c[j]*d
    ))
  ),
  return(fmod)
)$

/*
Solver
Solves the system of equation Kmod*u=fmod
Inputs:
  Kmod    Modified stiffness matrix
  fmod    Modified force matrix
Outputs:
  u_glob  Results (displacements)
*/
newsolver(Kmod, fmod) :=
block([u,i,u_solv,u_glob],
  u: zeromatrix(length(Kmod),1),
  for i: 1 thru length(Kmod) do (
      u[i][1]: concat('u_,i)
  ),
  /* establish the system of equations to solve*/
  eqns : Kmod.u - fmod,
  /* check for zero elements in "eqns" */
  for i:1 thru length(eqns) do(
    if (float(eqns[i][1])=0.0) then (
      eqns[i][1] : concat('u_,i)
  )),
  u_glob : zeromatrix(length(eqns),1),
  u_solv : solve(flatten(args(eqns)),flatten(args(u))),
  for i : 1 thru length(u_glob) do(
    u_glob[i][1] : second(u_solv[1][i])
  ),
  return(u_glob)
)$

/*
2D Planar Problem - Solution Driver
Inputs:
```

```
  nodxy   Nodal coordinates
  elenod  Element end nodes
  elemat  Element material properties
  elefab  Element fabrication properties
  nodtag  Nodal freedom tags list
  nodval  Nodal freedom values list
Outputs:
  noddis  Nodal displacements
  nodfor  Nodal forces
  elefor  Element forces
  elesig  Element stresses
*/
PlanarElementsSolution(nodxy, elenod, elemat, edges, pressures, body, nodtag,
     nodval, numerical) :=
block([K,Kmod,f,fmod,u,noddis,nodfor,elefor,elesig,eftab,Ke,i,c],
  /* calculate the global stiffness matrix */
  [K_els, K] : k_gl_2D_plane(nodxy, elenod, elemat, nen, type, numerical),
  /* calculate the modified global stiffness matrix */
  Kmod : ModifiedMasterStiffness(nodtag,K),
  /* calculate the global force matrix */
  if (nen = 4) then (
    f : f_edges_2D_plane_analytical(nodxy, pressures, length(nodxy), edges) )
  elseif (nen = 3) then (
    f : triangular_f_edges_2D_plane_analytical(nodxy, pressures,
     length(nodxy), edges)
  ),
  f : f + transpose(FlatNodePartVector(nodval)),
  f : f + body_gl_2D_Plane(nodxy, elenod, nen, body),
  /* calculate the modified global force matrix */
  fmod : ModifiedNodeForces(nodtag,nodval,K,f),
  /* solve the system of equation for displacements */
  u : invert(bfloat(Kmod)).fmod,
  /* transform u to noddis (distinguished nodal displacements) */
  noddis : zeromatrix(length(u)/2,2),
  c : 0,
  for i: 1 thru (length(u)/2) do(
    c : c + 1,
    noddis[i] : args(append(u[(c-1)*2+1],u[(c-1)*2+2]))
  ),
  return([noddis,Kmod,fmod])
)$

/*
Creates polynomial functions
Inputs:
  xi      coordinate values
  nen     Number of element nodes
Outputs:
  A       A array
  CHI     CHI array
*/
polynomial_maker(nen, xi):=
block([i],
  a : makelist(0,nen),
  chi: makelist(0,nen),
  for i: 1 thru nen do (
    a[i] : concat('a_,i-1),
    if (i=1 and xi=0) then (
      chi[i] : 1 )
```

```
    else (
      chi[i]: (xi^(i-1)) )
  ),
  return([a,chi])
)$

/*
(Prescribed) Displacement DOF Tags
Inputs:
  nodtag  Nodal freedom tags list
Output:
  pdof    Prescribed displacement DOF tags list
*/
PrescDispDOFTags(nodtag) :=
block([j,n,numnod:length(nodtag),pdof:[],k:0,m],
  for n : 1 thru numnod do (
    m : length(nodtag[n]),
    for j : 1 thru m do (
      if (is(nodtag[n][j] > 0)) then pdof : append(pdof, [k+j])
    ),
    k : k + m
  ),
  return(pdof)
)$

/*
(Prescribed) Displacement DOF Values
Inputs:
  nodtag  Nodal freedom tags list
  nodval  Nodal freedom values list
Output:
  pval    Prescribed displacement DOF values
*/
PrescDispDOFValues(nodtag, nodval) :=
block([j,n,numnod:length(nodtag),pval:[],k:0,m],
  for n : 1 thru numnod do (
    m : length(nodtag[n]),
    for j : 1 thru m do (
      if (is(nodtag[n][j] > 0)) then pval : append(pval, [nodval[n][j]])
    ),
    k : k + m
  ),
  return(pval)
)$

/*
Gauss Quadrature
Inputs:
  nqp     Number of quadrature points
Output:
  qp      Quadrature points array
*/
quadrature_1d(nqp) :=
block([p,q,v,w,x,qp],
  qp : zeromatrix(nqp,2),
  if (nqp=1) then
   (qp[1,1] : 0,
    qp[1,2] : 2)
  elseif (nqp=2) then
```

```
    (qp[1,1] : -1/sqrt(3),
     qp[1,2] : 1,
     qp[2,1] : 1/sqrt(3),
     qp[2,2] : 1)
     elseif (nqp=3) then
    (qp[1,1] : -sqrt(6/10),
     qp[1,2] : 5/9,
     qp[2,1] : 0,
     qp[2,2] : 8/9,
     qp[3,1] : +sqrt(6/10),
     qp[3,2] : 5/9)
     elseif (nqp=4) then
    (qp[1,1] : -(1/35)*sqrt(525+(70*sqrt(30))),
     qp[1,2] :  (1/36)*(18-sqrt(30)),
     qp[2,1] : -(1/35)*sqrt(525-(70*sqrt(30))),
     qp[2,2] :  (1/36)*(18+sqrt(30)),
     qp[3,1] :  (1/35)*sqrt(525-(70*sqrt(30))),
     qp[3,2] :  (1/36)*(18+sqrt(30)),
     qp[4,1] :  (1/35)*sqrt(525+(70*sqrt(30))),
     qp[4,2] :  (1/36)*(18-sqrt(30)))
     elseif (nqp=5) then
    (qp[1,1] : -(1/21)*sqrt(245+(14*sqrt(70))),
     qp[1,2] :  (1/900)*(322-(13*sqrt(70))),
     qp[2,1] : -(1/21)*sqrt(245-(14*sqrt(70))),
     qp[2,2] :  (1/900)*(322+(13*sqrt(70))),
     qp[3,1] : 0,
     qp[3,2] : 128/225,
     qp[4,1] :  (1/21)*sqrt(245-(14*sqrt(70))),
     qp[4,2] :  (1/900)*(322+(13*sqrt(70))),
     qp[5,1] :  (1/21)*sqrt(245+(14*sqrt(70))),
     qp[5,2] :  (1/900)*(322-(13*sqrt(70))))
     else
     (
       /*
       adapted from rosettacode.org
       https://rosettacode.org/wiki/Numerical_integration/
       Gauss-Legendre_Quadrature#Maxima
       */
       bftorat:false, /* controls the conversion of bfloats to rational numbers
      */
       ratepsilon : 2.0e-14, /* the tolerance used in the conversion of
      floating point numbers to rational numbers */
       fpprec_backup : fpprec,
       fpprec : 32,
       p: expand(legendre_p(nqp, x)),
       q: expand(nqp/2*diff(p, x)*legendre_p(nqp - 1, x)),
       v: map(rhs, bfallroots(p)),
       w: map(lambda([z], 1/subst([x = z], q)), v),
       for i:1 thru nqp do (
         qp[i,1] : (v[i]),
         qp[i,2] : (w[i])
       ),
       fpprec : fpprec_backup
     ),
   return(qp)
)$


/*
```

```
Gauss Quadrature - 2D
Quadrilateral Elements

Inputs:
  nqp    Number of quadrature points
Output:
  qp      Quadrature points array
*/
quadrature_2d(nqp) :=
block([p,q,v,w,x,qp],
  qp : zeromatrix(nqp,3),
  if (nqp=1) then
    (qp[1,1] : 0.0,
     qp[1,2] : 0.0,
     qp[1,3] : 4.0)
  elseif (nqp=4) then
    (qp[1,1] : -1/sqrt(3),
     qp[1,2] : -1/sqrt(3),
     qp[1,3] : 1.0,
     qp[2,1] : +1/sqrt(3),
     qp[2,2] : -1/sqrt(3),
     qp[2,3] : 1.0,
     qp[3,1] : +1/sqrt(3),
     qp[3,2] : +1/sqrt(3),
     qp[3,3] : 1.0,
     qp[4,1] : -1/sqrt(3),
     qp[4,2] : +1/sqrt(3),
     qp[4,3] : 1.0)
  elseif (nqp=9) then
    (qp[1,1] : -sqrt(6/10),
     qp[1,2] : -sqrt(6/10),
     qp[1,3] : 25/81,
     qp[2,1] : 0.0,
     qp[2,2] : -sqrt(6/10),
     qp[2,3] : 40/81,
     qp[3,1] : +sqrt(6/10),
     qp[3,2] : -sqrt(6/10),
     qp[3,3] : 25/81,
     qp[4,1] : -sqrt(6/10),
     qp[4,2] : 0.0,
     qp[4,3] : 40/81,
     qp[5,1] : 0.0,
     qp[5,2] : 0.0,
     qp[5,3] : 64/81,
     qp[6,1] : +sqrt(6/10),
     qp[6,2] : 0.0,
     qp[6,3] : 40/81,
     qp[7,1] : -sqrt(6/10),
     qp[7,2] : +sqrt(6/10),
     qp[7,3] : 25/81,
     qp[8,1] : 0.0,
     qp[8,2] : +sqrt(6/10),
     qp[8,3] : 40/81,
     qp[9,1] : +sqrt(6/10),
     qp[9,2] : +sqrt(6/10),
     qp[9,3] : 25/81)
     /*
     elseif (nqp=16) then
    (aq  : rat(sqrt( (3/7) + (2/7)*sqrt(6/5))),
```

```
    aqw : rat((18 - sqrt(30))/36),
    bq  : rat(sqrt( (3/7) - (2/7)*sqrt(6/5))),
    bqw : rat((18 + sqrt(30))/36),
    qp[1,1] : -aq,
    qp[1,2] : -aq,
    qp[1,3] : aqw^2,
    qp[2,1] : -bq,
    qp[2,2] : -aq,
    qp[2,3] : aqw*bqw,
    qp[3,1] : bq,
    qp[3,2] : -aq,
    qp[3,3] : aqw*bqw,
    qp[4,1] : aq,
    qp[4,2] : -aq,
    qp[4,3] : aqw^2,
    qp[5,1] : -aq,
    qp[5,2] : -bq,
    qp[5,3] : aqw*bqw,
    qp[6,1] : -bq,
    qp[6,2] : -bq,
    qp[6,3] : bqw^2,
    qp[7,1] : bq,
    qp[7,2] : -bq,
    qp[7,3] : bqw^2,
    qp[8,1] : aq,
    qp[8,2] : -bq,
    qp[8,3] : aqw*bqw,
    qp[9,1] : -aq,
    qp[9,2] : bq,
    qp[9,3] : aqw*bqw,
    qp[10,1] : -bq,
    qp[10,2] : bq,
    qp[10,3] : bqw^2,
    qp[11,1] : bq,
    qp[11,2] : bq,
    qp[11,3] : bqw^2,
    qp[12,1] : aq,
    qp[12,2] : bq,
    qp[12,3] : aqw*bqw,
    qp[13,1] : -aq,
    qp[13,2] : aq,
    qp[13,3] : aqw^2,
    qp[14,1] : -bq,
    qp[14,2] : aq,
    qp[14,3] : aqw*bqw,
    qp[15,1] : bq,
    qp[15,2] : aq,
    qp[15,3] : aqw*bqw,
    qp[16,1] : aq,
    qp[16,2] : aq,
    qp[16,3] : aqw^2)
    */
    else(
    noqpd : sqrt(nqp),
    qp_temp1 : (quadrature_1d(noqpd)),
    l_temp : 1,
    for j1 : 1 thru noqpd do (
      for i1 : 1 thru noqpd do (
        qp[l_temp,1] : qp_temp1[i1,1],
```

```
        qp[l_temp,2] : qp_temp1[j1,1],
        qp[l_temp,3] : qp_temp1[i1,2]*qp_temp1[j1,2],
        l_temp : l_temp + 1
      )
    ) ),

  return(qp)
)$

/*
2D Reissner Mindlin Plate Problem - Solution Driver

Inputs:
  nodxy   Nodal coordinates
  elenod  Element end nodes
  elemat  Element material properties
  elefab  Element fabrication properties
  nodtag  Nodal freedom tags list
  nodval  Nodal freedom values list

Outputs:
  noddis  Nodal displacements
  nodfor  Nodal forces
  elefor  Element forces
  elesig  Element stresses

*/
ReissnerMindlinPlateElementsSolution(nodxy, elenod, elemat, nodtag, nodval,
     numerical) :=
block([K,Kmod,f,fmod,u,noddis,nodfor,elefor,elesig,eftab,Ke,i,c],
  /* calculate the global stiffness matrix */
  [K_els, K] : k_gl_2D_Reissner_Mindlin_plate(nodxy, elenod, elemat, nen,
     type, numerical),
  /* calculate the modified global stiffness matrix */
  Kmod : ModifiedMasterStiffness(nodtag,K),
  /* calculate the global force matrix */
  /*
  if (nen = 4) then (
    f : f_edges_2D_plane_analytical(nodxy, pressures, length(nodxy), edges) )
  elseif (nen = 3) then (
    f : triangular_f_edges_2D_plane_analytical(nodxy, pressures,
     length(nodxy), edges)
  ),
  */
  f : transpose(FlatNodePartVector(nodval)),
  f : f + ratsimp(body_gl_2D_Reissner_Mindlin_plate(nodxy, elenod, nen)),
  /* calculate the modified global force matrix */
  fmod : ModifiedNodeForces(nodtag,nodval,K,f),

  /* solve the system of equation for displacements */
  u : invert(Kmod).fmod,
  /* transform u to noddis (distinguished nodal displacements) */
  noddis : zeromatrix(length(u)/3,3),
  c : 0,
  for i: 1 thru (length(u)/3) do(
    c : c + 1,
    noddis[i] : args(append(u[(c-1)*3+1],u[(c-1)*3+2],u[(c-1)*3+3]))
  ),
  return([float(noddis),Kmod,fmod])
```

```
)$

/*
Shape Functions Routine (1D) - N nodes
Inputs:
  nen     Number of element nodes
  ncoor_temp   Nodal coordinates
  xi_val  Local coordinate value
Outputs:
  NNN     Shape functions
  dNN     First derivatives of the shape functions
  jac     Jacobian
*/
shape_1d_nNodes_rod(nen, ncoor_temp, xi_val):=
block([i,j,xi,u,AA_inv,chi_val,AA,chi,x1,y1,x2,y2,x21,y21,LL,L,
NNN,dNN,jac],
  u : makelist(0,nen),
  AA : zeromatrix(nen,nen),
  AA_inv : zeromatrix(nen,nen),
  for i: 1 thru nen do (
    xi : -1 + (i-1)*(2/(nen-1)),
    [a, chi] : polynomial_maker(nen,xi),
    u[i] : chi.transpose(a),
    for j:1 thru nen do(
      AA_inv[i][j] : coeff(u[i],a[j])
    )
  ),
  AA : invert(AA_inv),
  kill(xi),
  [a, chi] : polynomial_maker(nen,xi),
  chi_val : subst(xi_val,xi,chi),
  NNN : subst(xi_val,xi,chi_val.AA),
  dNN : subst(xi_val,xi,diff(chi.AA,xi)),
  /* jacobian */
  [[x1,y1],[x2,y2]] : [ncoor_temp[1],ncoor_temp[nen]],
  [x21,y21] : [x2-x1,y2-y1],
  LL : (x21^2+y21^2),
  L : sqrt(LL),
  jac : L/2,
  NNN : NNN,
  dNN : dNN/jac,
  return([NNN,dNN,jac])
)$

/*
2D linear shape function routine

Inputs:
point(2)      : coordinates of the point
                point[1] : xi-direction
                point[2] : eta-direction

ncoor[4][2] : nodal point coordinates
                ncoor[:][1] : x-direction
                ncoor[:][2] : y-direction

Output:
shp(3,4)      : shape functions array
                shp(1,:) : derivatives in the x-direction
```

```
                shp(2,:) : derivatives in the y-direction
                shp(3,:) : shape functions
*/
shape_functions_2d(point, ncoor) :=
block([],

  /* calculation of the shape function values on "point" */
  shp[3,1] : (1/4)*(1-point[1])*(1-point[2]),
  shp[3,2] : (1/4)*(1+point[1])*(1-point[2]),
  shp[3,3] : (1/4)*(1+point[1])*(1+point[2]),
  shp[3,4] : (1/4)*(1-point[1])*(1+point[2]),

  /* calculation of derivatives of shape function on "point" */

  /* local derivatives */
  /* wrt. xi */
  shp[1,1] : (1/4)*(-1+point[2]),
  shp[1,2] : (1/4)*(+1-point[2]),
  shp[1,3] : (1/4)*(+1+point[2]),
  shp[1,4] : (1/4)*(-1-point[2]),
  /* wrt. eta */
  shp[2,1] : (1/4)*(-1+point[1]),
  shp[2,2] : (1/4)*(-1-point[1]),
  shp[2,3] : (1/4)*(+1+point[1]),
  shp[2,4] : (1/4)*(+1-point[1]),

  dxdxi[1,1] : 0.0,
  dxdxi[2,1] : 0.0,
  dxdxi[1,2] : 0.0,
  dxdxi[2,2] : 0.0,

  /* geometrical derivatives */
  for i : 1 thru 4 do (
    dxdxi[1,1] : dxdxi[1,1] + shp[1,i]*ncoor[i][1],
    dxdxi[2,1] : dxdxi[2,1] + shp[1,i]*ncoor[i][2],
    dxdxi[1,2] : dxdxi[1,2] + shp[2,i]*ncoor[i][1],
    dxdxi[2,2] : dxdxi[2,2] + shp[2,i]*ncoor[i][2]
  ),

  /* the Jacobian */
  jac : dxdxi[1,1]*dxdxi[2,2] - dxdxi[2,1]*dxdxi[1,2],

  /* geometrical derivatives (inverse) */
  dxidx[1,1] : +(1/jac)*dxdxi[2,2],
  dxidx[1,2] : -(1/jac)*dxdxi[1,2],
  dxidx[2,1] : -(1/jac)*dxdxi[2,1],
  dxidx[2,2] : +(1/jac)*dxdxi[1,1],

  /* global derivatives of shape functions */
  /* to avoid introducing new variables for the global derivatives of the shape functions, we will
     replace the previously calculated local derivatives with the new ones. Note that since the
     calculation of shp[2,i] needs the local value of shp[1,i] (and in the same way, the calculation
     of shp[1,i] needs shp[2,i] in the local coordinates), we first store the value of global shp[1,i]
     in to the "temp" variable, and after calculating shp[2,i] in global coordinates we replace
```

```
     shp[1,i] with "temp". */

  for i : 1 thru 4 do (
    temp : shp[1,i]*dxidx[1,1] + shp[2,i]*dxidx[2,1],
    shp[2,i] : shp[1,i]*dxidx[1,2] + shp[2,i]*dxidx[2,2],
    shp[1,i] : temp
  ),
  /* return the shape function array and the jacobian */
  return([shp,jac])
)$

/*
2D linear shape function routine
for the classical plate

Inputs:
point(2)      : coordinates of the point
                point[1] : xi-direction
                point[2] : eta-direction

ncoor[4][2] : nodal point coordinates
                ncoor[:][1] : x-direction
                ncoor[:][2] : y-direction

Output:
shp(3,3,4)      : shape functions array
                shp(:,1,:) : u (DOF)
                shp(:,2,:) : phi_x (DOF)
                shp(:,3,:) : phi_y (DOF)
                shp(1,:,:) : first derivatives in the x-direction
                shp(2,:,:) : first derivatives in the y-direction
                shp(3,:,:) : shape functions

shp2s(3,3,4)    : second derivatives of shape functions
                shp2s(1,:,:) : d2(N) / d(x^2)
                shp2s(2,:,:) : d2(N) / d(y^2)
                shp2s(3,:,:) : d2(N) / (d(x) d(y))

*/
shape_functions_2d_Kirschhoff_plate(point, ncoor) :=
block([shp,shp2s_l,dxdxi,dxidx,dxidx2,shp2s,shp_l,jac],
  /* local coordinates of nodes */

  /* Node 1: */
  xi_l[1]  : -1,
  eta_l[1] : -1,
  /* Node 2: */
  xi_l[2]  : +1,
  eta_l[2] : -1,
  /* Node 3: */
  xi_l[3]  : +1,
  eta_l[3] : +1,
  /* Node 4: */
  xi_l[4]  : -1,
  eta_l[4] : +1,


  /* geometrical derivatives */
  /* calculation of the shape function values on "point" */
```

```
shp_l[3,1] : (1/4)*(1-point[1])*(1-point[2]),
shp_l[3,2] : (1/4)*(1+point[1])*(1-point[2]),
shp_l[3,3] : (1/4)*(1+point[1])*(1+point[2]),
shp_l[3,4] : (1/4)*(1-point[1])*(1+point[2]),

/* calculation of derivatives of shape function on "point" */

/* local derivatives */
/* wrt. xi */
shp_l[1,1] : (1/4)*(-1+point[2]),
shp_l[1,2] : (1/4)*(+1-point[2]),
shp_l[1,3] : (1/4)*(+1+point[2]),
shp_l[1,4] : (1/4)*(-1-point[2]),
/* wrt. eta */
shp_l[2,1] : (1/4)*(-1+point[1]),
shp_l[2,2] : (1/4)*(-1-point[1]),
shp_l[2,3] : (1/4)*(+1+point[1]),
shp_l[2,4] : (1/4)*(+1-point[1]),

dxdxi[1,1] : 0.0,
dxdxi[2,1] : 0.0,
dxdxi[1,2] : 0.0,
dxdxi[2,2] : 0.0,

for i : 1 thru 4 do (
  dxdxi[1,1] : dxdxi[1,1] + shp_l[1,i]*ncoor[i][1],
  dxdxi[2,1] : dxdxi[2,1] + shp_l[1,i]*ncoor[i][2],
  dxdxi[1,2] : dxdxi[1,2] + shp_l[2,i]*ncoor[i][1],
  dxdxi[2,2] : dxdxi[2,2] + shp_l[2,i]*ncoor[i][2]
),

/* the Jacobian */
jac : dxdxi[1,1]*dxdxi[2,2] - dxdxi[2,1]*dxdxi[1,2],

/* geometrical derivatives (inverse) */
dxidx[1,1] : +(1/jac)*dxdxi[2,2],
dxidx[1,2] : -(1/jac)*dxdxi[1,2],
dxidx[2,1] : -(1/jac)*dxdxi[2,1],
dxidx[2,2] : +(1/jac)*dxdxi[1,1],

/* calculation of the shape function values on "point" */

for i : 1 thru 4 do (
  shp[3,1,i] : (1/8)*(1+xi_p*xi_l[i])*(1+eta_p*eta_l[i])*(2+xi_p*xi_l[i]
  +eta_p*eta_l[i]-xi_p^2-eta_p^2),
  shp[3,2,i] : (dxidx[2,2]^-1)/8*eta_l[i]*(1+xi_p*xi_l[i])*(eta_p*eta_l[i]
  -1)*(1+eta_p*eta_l[i])^2,
  shp[3,3,i] : -(dxidx[1,1]^-1)/8*xi_l[i]*(xi_p*xi_l[i]-1)*(1+eta_p
  *eta_l[i])*(1+xi_p*xi_l[i])^2
),

/* calculation of derivatives of shape function on "point" */

/* local derivatives */
/* wrt. xi */
for i : 1 thru 4 do ( /* loop over nodes */
  for j : 1 thru 3 do ( /* loop over degrees of freedom */
  /* d(N_u) / d(xi) */
```

```
      shp[1,j,i] :
       at(rat(diff(shp[3,j,i],xi_p,1)),[xi_p=point[1],eta_p=point[2]]),
      /* d(N_u) / d(eta) */
      shp[2,j,i] :
       at(rat(diff(shp[3,j,i],eta_p,1)),[xi_p=point[1],eta_p=point[2]]),
      /* d2(N_u) / d(xi^2) */
      shp2s_l[1,j,i] :
       at(rat(diff(shp[3,j,i],xi_p,2)),[xi_p=point[1],eta_p=point[2]]),
      /* d2(N_u) / d(eta^2) */
      shp2s_l[2,j,i] :
       at(rat(diff(shp[3,j,i],eta_p,2)),[xi_p=point[1],eta_p=point[2]]),
      /* d2(N_u) / (d(xi) d(eta)) */
      shp2s_l[3,j,i] :
       at(rat(diff(diff(shp[3,j,i],eta_p,1),xi_p,1)),[xi_p=point[1],
      eta_p=point[2]])
      )
    ),

    /* calculation of the shape functions based on the quadrature point
       coordinates */
    for i : 1 thru 4 do ( /* loop over nodes */
      for j : 1 thru 3 do ( /* loop over degrees of freedom */
        shp[3,j,i] : at(rat(shp[3,j,i]),[xi_p=point[1],eta_p=point[2]])
      )
    ),


    ddxidxdxi[1,1] : 0.0, /* d^2(xi)/(d(x)(dxi) */
    ddxidxdxi[1,2] : 0.0, /* d^2(xi)/(d(x)(deta) */
    ddxidxdxi[2,1] : 0.0, /* d^2(xi)/(d(y)(dxi) */
    ddxidxdxi[2,2] : 0.0, /* d^2(xi)/(d(y)(deta) */
    ddxidxdxi[3,1] : 0.0, /* d^2(eta)/(d(x)(dxi) */
    ddxidxdxi[3,2] : 0.0, /* d^2(eta)/(d(x)(deta) */
    ddxidxdxi[4,1] : 0.0, /* d^2(eta)/(d(y)(dxi) */
    ddxidxdxi[4,2] : 0.0, /* d^2(eta)/(d(y)(deta) */
    for i : 1 thru 4 do (
      for j : 1 thru 2 do (
        ddxidxdxi[1,j] : ddxidxdxi[1,j] + shp_l[j,i]*dxidx[j,1],
        ddxidxdxi[2,j] : ddxidxdxi[2,j] + shp_l[j,i]*dxidx[j,2],
        ddxidxdxi[3,j] : ddxidxdxi[3,j] + shp_l[j,i]*dxidx[j,1],
        ddxidxdxi[4,j] : ddxidxdxi[4,j] + shp_l[j,i]*dxidx[j,2]
      )
    ),
    /* Eqs. (4.53)-(4.58) */


    dxidx2[1,1] : ddxidxdxi[1,1]*dxidx[1,1] + ddxidxdxi[1,2]*dxidx[2,1],
    dxidx2[1,2] : ddxidxdxi[2,1]*dxidx[1,2] + ddxidxdxi[2,2]*dxidx[2,2],
    dxidx2[1,3] : ddxidxdxi[1,1]*dxidx[1,2] + ddxidxdxi[1,2]*dxidx[2,2],
    dxidx2[2,1] : ddxidxdxi[3,1]*dxidx[1,1] + ddxidxdxi[3,2]*dxidx[2,1],
    dxidx2[2,2] : ddxidxdxi[4,1]*dxidx[1,2] + ddxidxdxi[4,2]*dxidx[2,2],
    dxidx2[2,3] : ddxidxdxi[3,1]*dxidx[1,2] + ddxidxdxi[3,2]*dxidx[2,2],

    /* global derivatives of shape functions */
    /* to avoid introducing new variables for the global derivatives of the
       shape functions, we will
       replace the previously calculated local derivatives with the new ones.
       Note that since the
```

```
    calculation of shp[2,i] needs the local value of shp[1,i] (and in the same way, the calculation
    of shp[1,i] needs shp[2,i] in the local coordinates), we first store the value of global shp[1,i]
    in to the "temp" variable, and after calculating shp[2,i] in global coordinates we replace
    shp[1,i] with "temp". */

  for i : 1 thru 4 do ( /* loop over nodes */
    for j : 1 thru 3 do ( /* loop over degrees of freedom */
      shp2s[1,j,i] : factor(ratsimp(shp2s_l[1,j,i]*dxidx[1,1]^2 + shp2s_l[2,j,i]*dxidx[2,1]^2+2*shp2s_l[3,j,i]*dxidx[1,1]
      *dxidx[2,1]+shp[1,j,i]*dxidx2[1,1]+shp[2,j,i]*dxidx2[2,1])),
      shp2s[2,j,i] : factor(ratsimp(shp2s_l[1,j,i]*dxidx[1,2]^2 + shp2s_l[2,j,i]*dxidx[2,2]^2+2*shp2s_l[3,j,i]*dxidx[1,2]
      *dxidx[2,2]+shp[1,j,i]*dxidx2[1,2]+shp[2,j,i]*dxidx2[2,2])),
      shp2s[3,j,i] : factor(ratsimp(shp2s_l[1,j,i]*dxidx[1,1]*dxidx[1,2] + shp2s_l[2,j,i]*dxidx[2,1]*dxidx[2,2]+shp2s_l[3,j,i]*(dxidx[1,1]
      *dxidx[2,2]+dxidx[1,2]*dxidx[2,1])+shp[2,j,i]*dxidx2[1,3]
      +shp[2,j,i]*dxidx2[2,3]))
    )
  ),

  for i : 1 thru 4 do ( /* loop over nodes */
    for j : 1 thru 3 do ( /* loop over degrees of freedom */
      temp : factor(ratsimp(shp[1,j,i]*dxidx[1,1] + shp[2,j,i]*dxidx[2,1])),
      shp[2,j,i] : factor(ratsimp(shp[1,j,i]*dxidx[1,2] + shp[2,j,i]*dxidx[2,2])),
      shp[1,j,i] : temp
    )
  ),

  /* return the shape function array and the jacobian */
  return([(shp),(shp2s),(jac),(shp_l)])
)$

/*
2D linear elasticity element
plane stress and plane strain
strains on quadrature points
numerical integration

Inputs:
ncoor[4][2] : nodal point coordinates
              ncoor[:][1] : x-direction
              ncoor[:][2] : y-direction

nqp        : number of quadrature points per direction

nen         : number of nodes per element

displ       : nodal displacements

type        : type of the element
              "plane stress" or "plane strain"

Output:
strain      : strains on quadrature points

```

```
*/
strain_el_2D_plane_numerical(ncoor, nqp, nen, displ, type) :=
block([current_point, coef, i, l],

  /* calculate the quadrature points */
  qps : quadrature_2d(nqp),

  /* loop over the quadrature points */
  for l : 1 thru nqp do(

    /* initialize the elemental stiffness matrix */
    strain[l] : zeromatrix(3, 1),

    /* store the current quadrature point coordinates */
    current_point[1] : qps[l,1],
    current_point[2] : qps[l,2],

    /* calculate the shape functions and their derivatives at the current
     quadrature point*/
    [shape, jac] : shape_functions_2d(current_point, ncoor),

    /* calculation of B-matrix */
    B_mat : zeromatrix(8,3),
    for i : 1 thru nen do (
      B_mat[(i-1)*2+1,1] : shape[1,i],
      B_mat[(i-1)*2+1,3] : shape[2,i],
      B_mat[(i-1)*2+2,2] : shape[2,i],
      B_mat[(i-1)*2+2,3] : shape[1,i]
    ),

    /* calculation of BT_matrix */
    BT_mat : transpose(B_mat),

    /* calculation of strains on the current quadrature point */
    strain[l] : float(BT_mat.displ)

  ),

  /* return the strains on quadrature points */
  return(strain)
)$

/*
2D linear elasticity element
plane stress and plane strain
stresses on quadrature points

Inputs:
mat          : material properties

nqp          : number of quadrature points per direction

strains      : nodal displacements

type         : type of the element
               "plane stress" or "plane strain"

Output:
stress       : strains on quadrature points
```

```
*/
stress_el_2D_plane_numerical(mat, nqp, strain, type) :=
block([current_point, coef, i, l],

    /* calculation of the elasticity matrix */
  if (type = "plane stress") then (
    C_mat : zeromatrix(3,3),
    coef : mat[1]/(1-mat[2]^2),
    C_mat[1,1] : 1,
    C_mat[1,2] : mat[2],
    C_mat[2,1] : mat[2],
    C_mat[2,2] : 1,
    C_mat[3,3] : (1-mat[2])/2,
    C_mat : coef*C_mat
  ) elseif (type = "plane strain") then (
    C_mat : zeromatrix(3,3),
      coef : mat[1]/((1+mat[2])*(1-2*mat[2])),
      C_mat[1,1] : 1-mat[2],
      C_mat[1,2] : mat[2],
      C_mat[2,1] : mat[2],
      C_mat[2,2] : 1-mat[2],
      C_mat[3,3] : (1-2*mat[2])/2,
      C_mat : coef*C_mat
  ),

  /* loop over the quadrature points */
  for l : 1 thru nqp do(

    /* calculation of stresses on the current quadrature point */
    stress[l] : float(C_mat.strain[l])

  ),

  /* return the stresses on quadrature points */
  return(stress)
)$

/*
Symmetric Gauss Quadrature
Triangular Elements

Inputs:
  nqp    Number of quadrature points
Output:
  qp      Quadrature points array
*/
symmetric_quadrature_2d(nqp) :=
block([p,q,v,w,x,qp],
  qp : zeromatrix(nqp,3),
  if (nqp=1) then
   (qp[1,1] : 1/3,
    qp[1,2] : 1/3,
    qp[1,3] : 1/2)
  elseif (nqp=3) then
   (qp[1,1] : 1/6,
    qp[1,2] : 1/6,
    qp[1,3] : 1/6,
    qp[2,1] : 2/3,
```

```
      qp[2,2] : 1/6,
      qp[2,3] : 1/6,
      qp[3,1] : 1/6,
      qp[3,2] : 2/3,
      qp[3,3] : 1/6)
    elseif (nqp=4) then
     (qp[1,1] : 1/3,
      qp[1,2] : 1/3,
      qp[1,3] : -9/32,
      qp[2,1] : 3/5,
      qp[2,2] : 1/5,
      qp[2,3] : 25/96,
      qp[3,1] : 1/5,
      qp[3,2] : 3/5,
      qp[3,3] : 25/96,
      qp[4,1] : 1/5,
      qp[4,2] : 1/5,
      qp[4,3] : 25/96)
    elseif (nqp=7) then
     (qp[1,1] : 0.0,
      qp[1,2] : 0.0,
      qp[1,3] : 1/40,
      qp[2,1] : 1/2,
      qp[2,2] : 0.0,
      qp[2,3] : 1/15,
      qp[3,1] : 1.0,
      qp[3,2] : 0.0,
      qp[3,3] : 1/40,
      qp[4,1] : 1/2,
      qp[4,2] : 1/2,
      qp[4,3] : 1/15,
      qp[5,1] : 0.0,
      qp[5,2] : 1.0,
      qp[5,3] : 1/40,
      qp[6,1] : 0.0,
      qp[6,2] : 1/2,
      qp[6,3] : 1/15,
      qp[7,1] : 1/3,
      qp[7,2] : 1/3,
      qp[7,3] : 9/40)
    else(
      print("nqp value not supported!")
    ),
    return(qp)
)$

/*
2D linear elasticity element
elemental force matrix
pressure based
analytical integration
triangular element

Inputs:
ncoor(2,3) : nodal point coordinates
           ncoor[1,:] : x-direction
           ncoor[2,:] : y-direction

nen        : number of nodes per element
```

```
pressures  : pressures in the x- and y-directions
Output:

f_el : elemental force matrix

*/
triangular_f_edges_2D_plane_analytical(ncoor, pressures_e, nen, edges_e) :=
block([],

  /* initialize the elemental stiffness matrix */
  f_el : zeromatrix(2*nen, 1),

  /* loop over all edges */
  for e : 1 thru length(edges_e) do (
    /* loop over all nodal pairs of the edges */
    for n : 1 thru (length(edges[e])-1) do (
      Px_e[1] : pressures_e[e][1][1],
      Py_e[1] : pressures_e[e][1][2],
      Px_e[2] : pressures_e[e][2][1],
      Py_e[2] : pressures_e[e][2][2],

      n1 : edges_e[e][n],
      n2 : edges_e[e][n+1],
      ncoor_temp : [ncoor[n1], ncoor[n2]],

      /* calculate the shape functions and their derivatives at the quadrature
     point*/
      shape_temp[1,1] : xi,
      shape_temp[1,2] : 1.0 - xi,
      jac_temp : sqrt((ncoor_temp[2][2]-ncoor_temp[1][2])^2 +
     (ncoor_temp[2][1]-ncoor_temp[1][1])^2),

      Px : 0.0,
      Py : 0.0,
      for i : 1 thru 2 do (
        Px : Px + shape_temp[1,i]*Px_e[i],
        Py : Py + shape_temp[1,i]*Py_e[i]
      ),

      /* calcualtion of N_temp */

      /* define an array to store N matrices temporarily */
      N_temp : zeromatrix(nen*2,1),

      N_temp[n1*2 - 1,1] : shape_temp[1,1]*Px,
      N_temp[n1*2,1] : shape_temp[1,1]*Py,
      N_temp[n2*2 - 1,1] : shape_temp[1,2]*Px,
      N_temp[n2*2,1] : shape_temp[1,2]*Py,

      /* calculation of thickness on the current edge */
      current_point_x : 0,
      current_point_y : 0,
      for i : 1 thru 2 do(
        current_point_x : current_point_x + ncoor_temp[i][1]*shape_temp[1,i],
        current_point_y : current_point_y + ncoor_temp[i][2]*shape_temp[1,i]
      ),
      th : thickness(current_point_x,current_point_y),
```

```
      f_el : f_el + integrate(th*(N_temp)*(jac_temp),xi,0,1)

    )
  ),
  /* return the elemental force matrix */
  return(f_el)
)$

/*
2D linear shape function routine
(Triangular element)

Inputs:
point(2)     : coordinates of the point
               point[1] : xi-direction
               point[2] : eta-direction

ncoor[3][2] : nodal point coordinates
               ncoor[:][1] : x-direction
               ncoor[:][2] : y-direction

Output:
shp(3,3)     : shape functions array
               shp(1,:) : derivatives in the x-direction
               shp(2,:) : derivatives in the y-direction
               shp(3,:) : shape functions
*/
triangular_shape_functions_2d(point, ncoor) :=
block([],

  /* calculation of the shape function values on "point" */
  shp[3,1] : point[1],
  shp[3,2] : point[2],
  shp[3,3] : 1.0 - point[1] - point[2],

  /* calculation of derivatives of shape function on "point" */

  /* local derivatives */
  /* wrt. xi */
  shp[1,1] :  1.0,
  shp[1,2] :  0.0,
  shp[1,3] : -1.0,
  /* wrt. eta */
  shp[2,1] :  0.0,
  shp[2,2] :  1.0,
  shp[2,3] : -1.0,

  dxdxi[1,1] : 0.0,
  dxdxi[2,1] : 0.0,
  dxdxi[1,2] : 0.0,
  dxdxi[2,2] : 0.0,

  /* geometrical derivatives */
  for i : 1 thru 3 do (
    dxdxi[1,1] : dxdxi[1,1] + shp[1,i]*ncoor[i][1],
    dxdxi[2,1] : dxdxi[2,1] + shp[1,i]*ncoor[i][2],
    dxdxi[1,2] : dxdxi[1,2] + shp[2,i]*ncoor[i][1],
    dxdxi[2,2] : dxdxi[2,2] + shp[2,i]*ncoor[i][2]
  ),
```

```
  /* the Jacobian */
  jac : dxdxi[1,1]*dxdxi[2,2] - dxdxi[2,1]*dxdxi[1,2],

  /* geometrical derivatives (inverse) */
  dxidx[1,1] : +(1/jac)*dxdxi[2,2],
  dxidx[1,2] : -(1/jac)*dxdxi[1,2],
  dxidx[2,1] : -(1/jac)*dxdxi[2,1],
  dxidx[2,2] : +(1/jac)*dxdxi[1,1],

  /* global derivatives of shape functions */
  /* to avoid introducing new variables for the global derivatives of the
     shape functions, we will
     replace the previously calculated local derivatives with the new ones.
     Note that since the
     calculation of shp[2,i] needs the local value of shp[1,i] (and in the
     same way, the calculation
     of shp[1,i] needs shp[2,i] in the local coordinates), we first store the
     value of global shp[1,i]
     in to the "temp" variable, and after calculating shp[2,i] in global
     coordinates we replace
     shp[1,i] with "temp". */

  for i : 1 thru 3 do (
    temp : shp[1,i]*dxidx[1,1] + shp[2,i]*dxidx[2,1],
    shp[2,i] : shp[1,i]*dxidx[1,2] + shp[2,i]*dxidx[2,2],
    shp[1,i] : temp
  ),
  /* return the shape function array and the jacobian */
  return([shp,jac])
)$

/* end of file */
```

7.2 Graphical User Interface

The following file main.py is the Python source code for the graphical user interface. In addition, there is a ASCII configuration file (config.txt) required which contains the paths to the Maxima executable, e.g. C:\maxima-5.44.0\ bin\ wxmaxima.exe. The Python code was delopped in the programming environment wxPython [1] under consideration of particular libraries.

To easier navigate in the source code of the file main.py, Table 7.2 collects all the methods/functions and the corresponding line in the source code.

Table 7.2 List of all methods/functions in the GUI source code `main.py`

Class	Method/Function	Program line
TabOne		9
	__init__	10
	SaveDataBox	56
	OnSaveDataBrowse	74
	CreateMaterialGrid	83
	UpdateMaterialGrid	109
	UpdateMaterialGridColLabels	132
	ShowProperties	137
	SaveData	157
	LoadData	251
TabTwo		364
	__init__	365
	CreateNodeGrid	397
	CreateElementGrid	417
	AddMeshEdges	435
	CreateMeshGrid	447
	SetMeshGridNedge	466
	SetNodeGridNumnp	489
	SetElementGridNumel	516
	SetElementGridNen	536
	UpdateNodeGridRowLabels	556
	UpdateElementGridRowLabels	561
	UpdateElementGridColLabels	566
	UpdateMeshGridRowLabels	572
TabThree		578
	__init__	579
	CreateNodeTagGrid	608
	CreateNodeValGrid	626
	SetNodeTagValGridNdof	644
	AddPressureEdges	667
	CreatePressureEdgeGrid	676
	SetPressureGridNedge	695
	UpdateNodeTagGridRowLabels	716
	UpdateNodeTagGridColLabels	721
	UpdateNodeValGridRowLabels	726
	UpdateNodeValGridColLabels	731
TabFour		737
	__init__	790
	OnSaveBrowse	802
	OpenBox	811
	OnOpenBrowse	823
	ExportData	834
	ExportAndOpenData	1009

(continued)

Table 7.2 (continued)

Class	Method/Function	Program line
Problem		1018
	__init__	1019
	get_solver	1055
	GetProblemDescriptionBox	1071
	SetBody	1212
	SetBodyX	1217
	SetBodyY	1222
	SetKs	1227
	SetMaterialPropertiesPoisson	1232
	SetMaterialPropertiesShearModulus	1237
	SetMaterialPropertiesYoung	1242
	SetMx	1247
	SetMy	1252
	SetNdof	1257
	SetNdofNoEvent	1267
	SetNedge	1274
	SetNen	1280
	SetNmat	1286
	SetNqp	1293
	SetNumel	1298
	SetNumnp	1304
	SetQz	1310
	SetThickness	1315
MainFrame		1321
	__init__	1322
	SetStatusBar	1368
	SetVal	1378

main.py

```python
import os

import wx
import wx.grid
import simplejson as json


# Tab 1 - Problem Description
class TabOne(wx.Panel):
    def __init__(self, parent, ProblemList, MethodsList):
        wx.Panel.__init__(self, parent)
        global RadioBox1, RadioBox2, VerticalBox

        # Create a text control field for the JSON data files (to save/load problem settings)
        self.FDataPath = wx.TextCtrl(self, -1, size=(500, -1))
        self.saveDataButton = wx.Button(self, label="Save Data")
        self.loadDataButton = wx.Button(self, label="Load Data")
```

```

        # Create two radio boxes to choose the problem type and the solution method
        RadioBox1 = wx.RadioBox(self, label="Choose the problem type:", choices=ProblemList, style=wx.RA_SPECIFY_ROWS)
        RadioBox2 = wx.RadioBox(self, label="Choose the method:", choices=MethodsList, style=wx.RA_SPECIFY_ROWS)

        # Create a box to place the save/load data controls
        SaveVerticalBox = wx.StaticBox(self, -1, "Save and Load Data:")
        SaveVerticalBoxSizer = wx.StaticBoxSizer(SaveVerticalBox, wx.VERTICAL)
        SaveHorizontalBox = wx.BoxSizer(wx.HORIZONTAL)

        # Create a box to place the first tab controls
        VerticalBox = wx.BoxSizer(wx.VERTICAL)
        HorizontalBox = wx.BoxSizer(wx.HORIZONTAL)

        # Place and align different items to the created boxes (vertically and horizontally)
        VerticalBox.AddSpacer(10)
        VerticalBox.Add(HorizontalBox)
        HorizontalBox.Add(RadioBox1)
        HorizontalBox.AddSpacer(70)
        HorizontalBox.Add(RadioBox2)
        HorizontalBox.AddSpacer(70)
        SaveHorizontalBox.Add(self.loadDataButton)
        SaveHorizontalBox.AddSpacer(10)
        SaveHorizontalBox.Add(self.saveDataButton)
        SaveVerticalBoxSizer.Add(self.SaveDataBox())
        SaveVerticalBoxSizer.AddSpacer(10)
        SaveVerticalBoxSizer.Add(SaveHorizontalBox)
        HorizontalBox.Add(SaveVerticalBoxSizer)
        VerticalBox.AddSpacer(20)

        # Event handlers
        self.saveDataButton.Bind(wx.EVT_BUTTON, self.SaveData)
        self.loadDataButton.Bind(wx.EVT_BUTTON, self.LoadData)

        # Set the default sizer for the first tab
        self.SetSizer(VerticalBox)

    # Save/Load Data - Tab 1
    def SaveDataBox(self):
        # Create a local horizontal box
        HorizontalBox = wx.BoxSizer(wx.HORIZONTAL)

        # Create a browse button for the JSON data files (to save/load problem settings)
        BrowseButton = wx.Button(self, label="Browse")

        # Event handler
        BrowseButton.Bind(wx.EVT_BUTTON, self.OnSaveDataBrowse)

        # Alignments
        HorizontalBox.Add(self.FDataPath)
        HorizontalBox.AddSpacer(5)
        HorizontalBox.Add(BrowseButton)

        return HorizontalBox
```

```
    # Event - Save/Load Data - When the user click on Browse
    def OnSaveDataBrowse(self, event):
        with wx.FileDialog(self, "Save data file", wildcard="json files
     (*.json)|*.json",
                                  style=wx.FD_OPEN) as FileDialog:
            if FileDialog.ShowModal() == wx.ID_CANCEL:
                return  # the user changed their mind

            self.FDataPath.SetValue(FileDialog.GetPath())

    # Create a material grid to input different sets of material properties
    def CreateMaterialGrid(self):

        if Problem.Type == "Shear Deformable Plate":
            # Young's modulus, Poisson's ratio, and shear modulus
            NOfRows = 3
        else:
            # Young's modulus, Poisson's ratio
            NOfRows = 2
        NOfCols = 0

        MaterialGrid = wx.grid.Grid(self, style=wx.EXPAND)

        MaterialGrid.row = MaterialGrid.Col = 0
        MaterialGrid.CreateGrid(NOfRows, NOfCols)

        MaterialGrid.SetColLabelSize(30)
        MaterialGrid.SetRowLabelSize(120)

        MaterialGrid.SetRowLabelValue(0, "Young's Modulus")
        MaterialGrid.SetRowLabelValue(1, "Poisson's Ratio")
        if Problem.Type == "Shear Deformable Plate":
            MaterialGrid.SetRowLabelValue(2, "Shear Modulus")

        return MaterialGrid

    # Event - Update the material grid when the user change the number of
     material sets
    def UpdateMaterialGrid(self, event, MaterialGrid):

        Object = event.GetEventObject()

        if Object.GetValue() != "":

            CurrentCols = MaterialGrid.GetNumberCols()
            NewCols = int(Object.GetValue())

            if NewCols < CurrentCols:
                # delete rows
                MaterialGrid.DeleteCols(NewCols, CurrentCols, True)
            elif CurrentCols < NewCols:
                # add rows
                MaterialGrid.AppendCols(NewCols - CurrentCols)

            self.UpdateMaterialGridColLabels(MaterialGrid)

            self.Sizer.Layout()
```

```
            event.Skip()

    # Update the material grid column labels
    def UpdateMaterialGridColLabels(self, Grid):
        for Col in range(Grid.GetNumberCols()):
            Grid.SetColLabelValue(Col, "Material " + str(Col + 1))

    # Show the properties corresponding to the active problem type and solution method
    def ShowProperties(self):
        VerticalBox0 = wx.BoxSizer(wx.VERTICAL)

        VerticalBox0.AddSpacer(20)
        VerticalBox0.Add(Problem.GetProblemDescriptionBox())

        # Hide and detach the boxer related to the previously chosen problem
        VerticalBox.Hide(VerticalBox.GetItemCount() - 1)
        VerticalBox.Detach(VerticalBox.GetItemCount() - 1)

        # Add the new box
        VerticalBox.Add(VerticalBox0)

        # Update the layouts
        VerticalBox.Layout()
        self.Layout()

        return VerticalBox0

    # Event - Saves the data corresponding to the problem in a previously chosen file path
    def SaveData(self, event):
        DataFile = open(self.FDataPath.GetValue(), "w+")

        Data = {}
        Data["type"] = Problem.Type
        Data["method"] = Problem.Method
        Data["nMat"] = Problem.NMat
        MaterialsList = []
        for j in range(Problem.NMat):
            if Problem.Type == "Shear Deformable Plate":
                for i in range(3):
                    MaterialsList.append(MaterialGrid.GetCellValue(i, j))
            else:
                for i in range(2):
                    MaterialsList.append(MaterialGrid.GetCellValue(i, j))

        Data["material_data"] = MaterialsList
        Data["thickness"] = Problem.Thickness

        if Problem.Type == "Shear Deformable Plate":
            Data["shearcf"] = Problem.ShearCF

        Data["numnp"] = Problem.NUMNP
        Data["numel"] = Problem.NUMEL
        Data["nen"] = Problem.NEN

        if Problem.Method == "Numerical":
            Data["nqp"] = Problem.NQP

```

```
        if Problem.Type == "Plane Stress" or Problem.Type == "Plane Strain":
            Data["nedge"] = Problem.NEdge
            Data["bodyx"] = Problem.BodyX
            Data["bodyy"] = Problem.BodyY
        elif Problem.Type == "Classical Plate":
            Data["body"] = Problem.Body
        elif Problem.Type == "Shear Deformable Plate":
            Data["qz"] = Problem.QZ
            Data["mx"] = Problem.MX
            Data["my"] = Problem.MY

        Coordinates = []
        for i in range(Problem.NUMNP):
            for j in range(2):
                Coordinates.append(Tab2.NodeGrid.GetCellValue(i, j))
        Data["coordinates"] = Coordinates

        Connectivity = []
        for i in range(Problem.NUMEL):
            Connectivity.append(Tab2.ElementGrid.GetCellValue(i, 0))
            for j in range(1, Problem.NEN + 1):
                Connectivity.append(Tab2.ElementGrid.GetCellValue(i, j))
        Data["connectivity"] = Connectivity

        Edges = []
        Pressures = []
        Counter = 0
        for i in range(Problem.NEdge):
            Edges.append(Tab2.EdgeGrid.GetCellValue(i, 0))
            Pressures.append(Tab3.PressureGrid.GetCellValue(i, 0))
            Counter += 1
        Data["edges"] = Edges
        Data["pressures"] = Pressures

        BCTags = []
        Counter = 0
        for i in range(Problem.NUMNP):
            if Problem.Type == "Plane Stress" or Problem.Type == "Plane
    Strain":
                for j in range(2):
                    BCTags.append(Tab3.NodeTagGrid.GetCellValue(i, j))
                    Counter += 1
            elif Problem.Type == "Classical Plate" or Problem.Type == "Shear
    Deformable Plate":
                for j in range(3):
                    BCTags.append(Tab3.NodeTagGrid.GetCellValue(i, j))
                    Counter += 1
        Data["bctags"] = BCTags

        BCVals = []
        Counter = 0
        for i in range(Problem.NUMNP):
            if Problem.Type == "Plane Stress" or Problem.Type == "Plane
    Strain":
                for j in range(2):
                    BCVals.append(Tab3.NodeValGrid.GetCellValue(i, j))
                    Counter += 1
            elif Problem.Type == "Classical Plate" or Problem.Type == "Shear
    Deformable Plate":
```

```
                for j in range(3):
                    BCVals.append(Tab3.NodeValGrid.GetCellValue(i, j))
                    Counter += 1
        Data["bcvals"] = BCVals

        Data["outputfile"] = Tab4.FSavePath.GetValue()

        DataFile.write(json.dumps(Data, indent=2, sort_keys=True))

    # Event - Loads the data corresponding to the problem from a previously
     chosen file path
    def LoadData(self, event):
        DataFile = open(self.FDataPath.GetValue(), "r+")

        Data = json.loads(DataFile.read())

        Type = RadioBox1.FindString(Data["type"], bCase=True)
        RadioBox1.SetSelection(Type)
        Problem.Type = Data["type"]

        Method = RadioBox2.FindString(Data["method"], bCase=True)
        RadioBox2.SetSelection(Method)
        Problem.Method = Data["method"]

        Tab1.ShowProperties()

        MainFrame.SetStatusBar()

        Problem.NMatBox.SetValue(str(Data["nMat"]))

        MaterialsList = Data["material_data"]
        Counter = 0
        for j in range(Problem.NMat):
            if Problem.Type == "Shear Deformable Plate":
                for i in range(3):
                    MaterialGrid.SetCellValue(i, j, MaterialsList[Counter])
                    Counter += 1
            else:
                for i in range(2):
                    MaterialGrid.SetCellValue(i, j, MaterialsList[Counter])
                    Counter += 1

        Problem.ThicknessBox.SetValue(Data["thickness"])

        if Problem.Type == "Shear Deformable Plate":
            Problem.ShearCFBox.SetValue(Data["shearcf"])
            Problem.SetNdofNoEvent()

        Problem.NUMNPBox.SetValue(str(Data["numnp"]))
        Problem.NUMELBox.SetValue(str(Data["numel"]))
        Problem.NENBox.SetValue(str(Data["nen"]))

        if Problem.Method == "Numerical":
            Problem.NQPBox.SetValue(str(Data["nqp"]))

        if Problem.Type == "Plane Stress" or Problem.Type == "Plane Strain":
            # initialize edge and pressure grids if PressureGrid does not
 exist (to avoid duplicates)
            if not (hasattr(Tab3, 'PressureGrid')):
```

```
                Tab2.AddMeshEdges(event)
                Tab3.AddPressureEdges(event)

            Problem.NEdgeBox.SetValue(str(Data["nedge"]))
            Problem.BodyXBox.SetValue(Data["bodyx"])
            Problem.BodyYBox.SetValue(Data["bodyy"])
        elif Problem.Type == "Classical Plate":
            Problem.BodyBox.SetValue(Data["body"])
        elif Problem.Type == "Shear Deformable Plate":
            Problem.QZBox.SetValue(Data["qz"])
            Problem.MXBox.SetValue(Data["mx"])
            Problem.MYBox.SetValue(Data["my"])

        coordinates = Data["coordinates"]
        Counter = 0
        for i in range(Problem.NUMNP):
            for j in range(2):
                Tab2.NodeGrid.SetCellValue(i, j, coordinates[Counter])
                Counter += 1

        Connectivity = Data["connectivity"]
        Counter = 0
        for i in range(Problem.NUMEL):
            Tab2.ElementGrid.SetCellValue(i, 0, Connectivity[Counter])
            Counter += 1
            for j in range(1, Problem.NEN + 1):
                Tab2.ElementGrid.SetCellValue(i, j, Connectivity[Counter])
                Counter += 1

        if Problem.Type == "Plane Stress" or Problem.Type == "Plane Strain":
            Edges = Data["edges"]
            Pressures = Data["pressures"]
            Counter = 0
            for i in range(Problem.NEdge):
                Tab2.EdgeGrid.SetCellValue(i, 0, Edges[Counter])
                Tab3.PressureGrid.SetCellValue(i, 0, Pressures[Counter])
                Counter += 1

        BCTags = Data["bctags"]
        Counter = 0
        for i in range(Problem.NUMNP):
            if Problem.Type == "Plane Stress" or Problem.Type == "Plane Strain":
                for j in range(2):
                    Tab3.NodeTagGrid.SetCellValue(i, j, BCTags[Counter])
                    Counter += 1
            elif Problem.Type == "Classical Plate" or Problem.Type == "Shear Deformable Plate":
                for j in range(3):
                    Tab3.NodeTagGrid.SetCellValue(i, j, BCTags[Counter])
                    Counter += 1

        BCVals = Data["bcvals"]
        Counter = 0
        for i in range(Problem.NUMNP):
            if Problem.Type == "Plane Stress" or Problem.Type == "Plane Strain":
                for j in range(2):
                    Tab3.NodeValGrid.SetCellValue(i, j, BCVals[Counter])
```

```
                    Counter += 1
            elif Problem.Type == "Classical Plate" or Problem.Type == "Shear Deformable Plate":
                for j in range(3):
                    Tab3.NodeValGrid.SetCellValue(i, j, BCVals[Counter])
                    Counter += 1

        Tab4.FSavePath.SetValue(Data["outputfile"])


# Tab 2 - Nodal Coordinates and Element Connectivity
class TabTwo(wx.Panel):
    def __init__(self, parent):
        wx.Panel.__init__(self, parent)

        # Create two separate grids for nodal coordinates and element connectivity data
        self.NodeGrid = self.CreateNodeGrid()
        self.ElementGrid = self.CreateElementGrid()

        self.HorizontalBox = wx.BoxSizer(wx.HORIZONTAL)

        NodeBox = wx.StaticBox(self, -1, "Nodal Coordinates:")
        NodeBoxSizer = wx.StaticBoxSizer(NodeBox, wx.VERTICAL)
        NodeBoxSizer.Add(self.NodeGrid)

        ElementBox = wx.StaticBox(self, -1, "Element Connectivity:")
        ElementBoxSizer = wx.StaticBoxSizer(ElementBox, wx.VERTICAL)
        ElementBoxSizer.Add(self.ElementGrid)
        ElementBoxSizer.AddSpacer(5)
        Text = wx.StaticText(self, label="*The numbering of element nodes should be Counter-clockwise.")
        ElementBoxSizer.Add(Text)

        self.HorizontalBox.Add(NodeBoxSizer)
        self.HorizontalBox.AddSpacer(30)
        self.HorizontalBox.Add(ElementBoxSizer)

        EdgeBox = wx.StaticBox(self, -1, "Mesh Edges:")
        self.EdgeBoxSizer = wx.StaticBoxSizer(EdgeBox, wx.VERTICAL)

        self.HorizontalBox.AddSpacer(30)
        self.HorizontalBox.Add(self.EdgeBoxSizer)
        self.SetSizer(self.HorizontalBox)

    # Initialize the nodal coordinates grid (the number of rows will then get updated based on NUMNP)
    def CreateNodeGrid(self):
        NOfRows = 1
        NOfCols = 2

        NodeGrid = wx.grid.Grid(self, style=wx.EXPAND)

        NodeGrid.row = NodeGrid.Col = 0
        NodeGrid.CreateGrid(NOfRows, NOfCols)

        NodeGrid.SetColLabelSize(30)
        NodeGrid.SetRowLabelSize(60)

```

```
        NodeGrid.SetColLabelValue(0, "X Coordinate")
        NodeGrid.SetColLabelValue(1, "Y Coordinate")

        self.UpdateNodeGridRowLabels(NodeGrid)

        return NodeGrid

    # Initialize the element connectivity grid (the number of rows will then get updated based on NUMEL)
    def CreateElementGrid(self):
        NOfRows = 1
        NOfCols = 5

        ElementGrid = wx.grid.Grid(self)

        ElementGrid.row = ElementGrid.Col = 0
        ElementGrid.CreateGrid(NOfRows, NOfCols)

        ElementGrid.SetColLabelSize(30)
        ElementGrid.SetRowLabelSize(90)

        self.UpdateElementGridRowLabels(ElementGrid)
        self.UpdateElementGridColLabels(ElementGrid)

        return ElementGrid

    # Event - If necessary, a grid for mesh edges will be added to Tab 2 (based on the problem settings)
    def AddMeshEdges(self, event):

        # Create a grid for mesh edges (mesh grid) and add it to Tab 2
        self.EdgeGrid = self.CreateMeshGrid()
        self.EdgeBoxSizer.Add(self.EdgeGrid)

        # Update the layouts
        self.EdgeBoxSizer.Layout()
        self.Sizer.Layout()

    # Initialize a mesh grid based on the nodes on the edges of the model
    # (the number of rows will then get updated based on NEDGE)
    def CreateMeshGrid(self):
        NOfRows = 1
        NOfCols = 1

        MeshGrid = wx.grid.Grid(self, wx.EXPAND)

        MeshGrid.row = MeshGrid.Col = 0
        MeshGrid.CreateGrid(NOfRows, NOfCols)

        MeshGrid.SetColLabelSize(40)
        MeshGrid.SetRowLabelSize(90)

        MeshGrid.SetColLabelValue(0, "Nodes, \n e.g., 1,2,3")

        self.UpdateMeshGridRowLabels(MeshGrid)

        return MeshGrid

    # Event - Update the mesh grid size when the user change the value of NEDGE
```

```
    def SetMeshGridNedge(self, event):

        Object = event.GetEventObject()

        if Object.GetValue() != "":

            CurrentRows = self.EdgeGrid.GetNumberRows()
            NewRows = int(Object.GetValue())

            if NewRows < CurrentRows:
                # delete rows
                self.EdgeGrid.DeleteRows(NewRows, CurrentRows, True)
            elif CurrentRows < NewRows:
                # add rows
                self.EdgeGrid.AppendRows(NewRows - CurrentRows)

            self.UpdateMeshGridRowLabels(self.EdgeGrid)

            self.Sizer.Layout()

        event.Skip()

    # Event - Update the nodal coordinates grid size when the user change the
     value of NUMNP
    def SetNodeGridNumnp(self, event):
        Object = event.GetEventObject()

        CurrentRows = self.NodeGrid.GetNumberRows()
        NewRows = int(Object.GetValue())

        if NewRows < CurrentRows:
            # delete rows
            self.NodeGrid.DeleteRows(NewRows, CurrentRows, True)
            Tab3.NodeTagGrid.DeleteRows(NewRows, CurrentRows, True)
            Tab3.NodeValGrid.DeleteRows(NewRows, CurrentRows, True)
        elif CurrentRows < NewRows:
            # add rows
            self.NodeGrid.AppendRows(NewRows - CurrentRows)
            Tab3.NodeTagGrid.AppendRows(NewRows - CurrentRows)
            Tab3.NodeValGrid.AppendRows(NewRows - CurrentRows)

        self.UpdateNodeGridRowLabels(self.NodeGrid)
        Tab3.UpdateNodeTagGridRowLabels(Tab3.NodeTagGrid)
        Tab3.UpdateNodeValGridRowLabels(Tab3.NodeValGrid)

        self.Sizer.Layout()
        Tab3.Sizer.Layout()

        event.Skip()

    # Event - Update the element connectivity grid size when the user change
     the value of NUMEL
    def SetElementGridNumel(self, event):
        Object = event.GetEventObject()

        CurrentRows = self.ElementGrid.GetNumberRows()
        NewRows = int(Object.GetValue())

        if NewRows < CurrentRows:
```

```
            # delete rows
            self.ElementGrid.DeleteRows(NewRows, CurrentRows, True)
        elif CurrentRows < NewRows:
            # add rows
            self.ElementGrid.AppendRows(NewRows - CurrentRows)

        self.UpdateElementGridRowLabels(self.ElementGrid)

        self.Sizer.Layout()

        event.Skip()

    # Event - Update the element connectivity grid size when the user change
     the value of NEN
    def SetElementGridNen(self, event):
        Object = event.GetEventObject()

        CurrentCols = self.ElementGrid.GetNumberCols()
        NewCols = int(Object.GetValue()) + 1

        if NewCols < CurrentCols:
            # delete Columns
            self.ElementGrid.DeleteCols(NewCols, CurrentCols, True)
        elif CurrentCols < NewCols:
            # add Columns
            self.ElementGrid.AppendCols(NewCols - CurrentCols)

        self.UpdateElementGridColLabels(self.ElementGrid)

        self.Sizer.Layout()

        event.Skip()

    # Updates the nodal coordinates grid row labels
    def UpdateNodeGridRowLabels(self, Grid):
        for row in range(Grid.GetNumberRows()):
            Grid.SetRowLabelValue(row, "Node " + str(row + 1))

    # Updates the element connectivity grid row labels
    def UpdateElementGridRowLabels(self, Grid):
        for row in range(Grid.GetNumberRows()):
            Grid.SetRowLabelValue(row, "Element " + str(row + 1))

    # Updates the element connectivity grid column labels
    def UpdateElementGridColLabels(self, Grid):
        Grid.SetColLabelValue(0, "Material Set")
        for Col in range(1, Grid.GetNumberCols()):
            Grid.SetColLabelValue(Col, "Node " + str(Col))

    # Updates the mesh grid row labels
    def UpdateMeshGridRowLabels(self, Grid):
        for row in range(Grid.GetNumberRows()):
            Grid.SetRowLabelValue(row, "Edge " + str(row + 1))


# Tab 3 - Boundary Conditions
class TabThree(wx.Panel):
    def __init__(self, parent):
        wx.Panel.__init__(self, parent)
```

```
        t = wx.StaticText(self, -1)

        # Create two separate grids for NodeTag and NodeVal arrays
        self.NodeTagGrid = self.CreateNodeTagGrid()
        self.NodeValGrid = self.CreateNodeValGrid()

        HorizontalBox = wx.BoxSizer(wx.HORIZONTAL)

        NodeTagBox = wx.StaticBox(self, -1, "Nodal BC Tags:")
        NodeTagBoxSizer = wx.StaticBoxSizer(NodeTagBox, wx.VERTICAL)
        NodeTagBoxSizer.Add(self.NodeTagGrid)

        NodeValBox = wx.StaticBox(self, -1, "Nodal BC Values:")
        NodeValBoxSizer = wx.StaticBoxSizer(NodeValBox, wx.VERTICAL)
        NodeValBoxSizer.Add(self.NodeValGrid)

        PressureBox = wx.StaticBox(self, -1, "Pressure on Edges:")
        self.PressureBoxSizer = wx.StaticBoxSizer(PressureBox, wx.VERTICAL)

        HorizontalBox.Add(NodeTagBoxSizer)
        HorizontalBox.AddSpacer(30)
        HorizontalBox.Add(NodeValBoxSizer)
        HorizontalBox.AddSpacer(30)
        HorizontalBox.Add(self.PressureBoxSizer)
        self.SetSizer(HorizontalBox)

    # Initialize the nodal BC tags grid (the number of rows will then get updated based on NUMNP)
    def CreateNodeTagGrid(self):
        NOfRows = Tab2.NodeGrid.GetNumberRows()
        NOfCols = Problem.NDOF

        NodeTagGrid = wx.grid.Grid(self, style=wx.EXPAND)

        NodeTagGrid.row = NodeTagGrid.Col = 0
        NodeTagGrid.CreateGrid(NOfRows, NOfCols)

        NodeTagGrid.SetColLabelSize(30)
        NodeTagGrid.SetRowLabelSize(60)

        self.UpdateNodeTagGridColLabels(NodeTagGrid)
        self.UpdateNodeTagGridRowLabels(NodeTagGrid)

        return NodeTagGrid

    # Initialize the nodal BC values grid (the number of rows will then get updated based on NUMNP)
    def CreateNodeValGrid(self):
        NOfRows = Tab2.NodeGrid.GetNumberRows()
        NOfCols = Problem.NDOF

        NodeValGrid = wx.grid.Grid(self)

        NodeValGrid.row = NodeValGrid.Col = 0
        NodeValGrid.CreateGrid(NOfRows, NOfCols)

        NodeValGrid.SetColLabelSize(30)
        NodeValGrid.SetRowLabelSize(90)
```

```
            self.UpdateNodeValGridRowLabels(NodeValGrid)
            self.UpdateNodeValGridColLabels(NodeValGrid)

            return NodeValGrid

        # Event - Update the NodeTag and NodeVal grids (based on the problem settings)
        def SetNodeTagValGridNdof(self, event):
            Object = event.GetEventObject()
            # change the Columns in Tab3 (NDOF)
            CurrentCols = self.NodeTagGrid.GetNumberCols()
            NewCols = Problem.NDOF
            # print(NewCols, CurrentCols)

            if NewCols < CurrentCols:
                # delete Cols
                self.NodeTagGrid.DeleteCols(NewCols, CurrentCols, True)
                self.NodeValGrid.DeleteCols(NewCols, CurrentCols, True)
            elif CurrentCols < NewCols:
                # add Cols
                self.NodeTagGrid.AppendCols(NewCols - CurrentCols)
                self.NodeValGrid.AppendCols(NewCols - CurrentCols)

            self.UpdateNodeTagGridColLabels(self.NodeTagGrid)
            self.UpdateNodeValGridColLabels(self.NodeValGrid)

            self.Sizer.Layout()
            event.Skip()

        # Event - If necessary, a grid for pressure edges will be added to Tab 3 (based on the problem settings)
        def AddPressureEdges(self, event):

            self.PressureGrid = self.CreatePressureEdgeGrid()
            self.PressureBoxSizer.Add(self.PressureGrid)
            self.PressureBoxSizer.Layout()
            self.Sizer.Layout()

        # Initialize a pressure edge grid based on the nodes on the edges of the model
        # (the number of rows will then get updated based on NEDGE)
        def CreatePressureEdgeGrid(self):
            NOfRows = 1
            NOfCols = 1

            PressureEdgeGrid = wx.grid.Grid(self, wx.EXPAND)

            PressureEdgeGrid.row = PressureEdgeGrid.Col = 0
            PressureEdgeGrid.CreateGrid(NOfRows, NOfCols)

            PressureEdgeGrid.SetColLabelSize(40)
            PressureEdgeGrid.SetRowLabelSize(90)

            PressureEdgeGrid.SetColLabelValue(0, "Pressure\np_x, p_y")

            Tab2.UpdateMeshGridRowLabels(PressureEdgeGrid)

            return PressureEdgeGrid

```

```
    # Event - Update the pressure edge grid size when the user change the
     value of NEDGE
    def SetPressureGridNedge(self, event):
        Object = event.GetEventObject()
        if Object.GetValue() != "":

            CurrentRows = self.PressureGrid.GetNumberRows()
            NewRows = int(Object.GetValue())

            if NewRows < CurrentRows:
                # delete rows
                self.PressureGrid.DeleteRows(NewRows, CurrentRows, True)
            elif CurrentRows < NewRows:
                # add rows
                self.PressureGrid.AppendRows(NewRows - CurrentRows)

            Tab2.UpdateMeshGridRowLabels(self.PressureGrid)

        Tab3.Sizer.Layout()

        event.Skip()

    # Updates the NodeTag grid row labels
    def UpdateNodeTagGridRowLabels(self, Grid):
        for row in range(Grid.GetNumberRows()):
            Grid.SetRowLabelValue(row, "Node " + str(row + 1))

    # Updates the NodeTag grid column labels
    def UpdateNodeTagGridColLabels(self, Grid):
        for Col in range(Grid.GetNumberCols()):
            Grid.SetColLabelValue(Col, "DOF " + str(Col + 1))

    # Updates the NodeVal grid row labels
    def UpdateNodeValGridRowLabels(self, Grid):
        for row in range(Grid.GetNumberRows()):
            Grid.SetRowLabelValue(row, "Node " + str(row + 1))

    # Updates the NodeTag grid column labels
    def UpdateNodeValGridColLabels(self, Grid):
        for Col in range(Grid.GetNumberCols()):
            Grid.SetColLabelValue(Col, "DOF " + str(Col + 1))


# Tab 4 - Output
class TabFour(wx.Panel):
    def __init__(self, parent):
        wx.Panel.__init__(self, parent)

        # Create export [/and open] buttons
        self.ExportButton = wx.Button(self, label="Export")
        self.ExportAndOpenButton = wx.Button(self, label="Export and open in
     wxMaxima")

        self.FSavePath = wx.TextCtrl(self, -1, size=(500, -1))
        self.FOpenPath = wx.TextCtrl(self, -1, size=(500, -1))

        # Save WxMaxima path in a config.txt file (the routine will load the
     path if "config.txt" exists)
```

```
        if os.path.isfile("./config.txt") and os.stat("./config.txt").st_size != 0:
            ConfigFile = open("./config.txt", "r+")
            OpenPath = ConfigFile.readline()
            print(OpenPath)
            self.FOpenPath.SetValue(OpenPath)

        ExportBoxer = wx.BoxSizer(wx.VERTICAL)

        OutputBox = wx.StaticBox(self, -1, "Choose the output file:")
        OutputBoxSizer = wx.StaticBoxSizer(OutputBox, wx.VERTICAL)

        wxmaximaBox = wx.StaticBox(self, -1, "Choose the wxMaxima executable file (optional):")
        WXMaximaBoxSizer = wx.StaticBoxSizer(wxmaximaBox, wx.VERTICAL)

        savebox = self.SaveBox()
        OutputBoxSizer.Add(savebox)
        OutputBoxSizer.AddSpacer(10)

        ExportBoxer.Add(OutputBoxSizer)
        ExportBoxer.AddSpacer(10)

        OpenBox = self.OpenBox()
        WXMaximaBoxSizer.Add(OpenBox)
        ExportBoxer.Add(WXMaximaBoxSizer)

        ExportButtonSizer = wx.BoxSizer(wx.HORIZONTAL)

        ExportButtonSizer.Add(self.ExportButton)
        ExportButtonSizer.AddSpacer(10)
        ExportButtonSizer.Add(self.ExportAndOpenButton)

        ExportBoxer.AddSpacer(15)
        ExportBoxer.Add(ExportButtonSizer)

        self.SetSizer(ExportBoxer)

        # Event handlers
        self.ExportButton.Bind(wx.EVT_BUTTON, self.ExportData)
        self.ExportAndOpenButton.Bind(wx.EVT_BUTTON, self.ExportAndOpenData)

    # Save Data (wxMaxima input file) - Tab 4
    def SaveBox(self):
        HorizontalBox = wx.BoxSizer(wx.HORIZONTAL)

        BrowseButton = wx.Button(self, label="Browse")

        BrowseButton.Bind(wx.EVT_BUTTON, self.OnSaveBrowse)
        HorizontalBox.Add(self.FSavePath)
        HorizontalBox.AddSpacer(5)
        HorizontalBox.Add(BrowseButton)
        return HorizontalBox

    # Event - Save Data - When the user click on Browse
    def OnSaveBrowse(self, event):
        with wx.FileDialog(self, "Save wxMaxima file", wildcard="WXM files (*.wxm)|*.wxm",
```

```
                              style=wx.FD_SAVE | wx.FD_OVERWRITE_PROMPT) as
 FileDialog:
            if FileDialog.ShowModal() == wx.ID_CANCEL:
                return  # the user changed their mind

            self.FSavePath.SetValue(FileDialog.GetPath())

    # Open (WxMaxima executable file) - Tab 4
    def OpenBox(self):
        HorizontalBox = wx.BoxSizer(wx.HORIZONTAL)

        BrowseButton = wx.Button(self, label="Browse")

        BrowseButton.Bind(wx.EVT_BUTTON, self.OnOpenBrowse)
        HorizontalBox.Add(self.FOpenPath)
        HorizontalBox.AddSpacer(5)
        HorizontalBox.Add(BrowseButton)
        return HorizontalBox

    # Event - Open (wxMaxima executable file) - When the user click on Browse
    def OnOpenBrowse(self, event):
        with wx.FileDialog(self, "Open wxMaxima executable",
 wildcard="wxmaxima.exe|wxmaxima.exe",
                              style=wx.FD_OPEN) as FileDialog:
            if FileDialog.ShowModal() == wx.ID_CANCEL:
                return  # the user changed their mind

            self.FOpenPath.SetValue(FileDialog.GetPath())
            ConfigFile = open("config.txt", "w+")
            ConfigFile.write(FileDialog.GetPath())

    # Event  - Save Data (wxMaxima input file) - When the user click on Export
 (or Export and open...)
    def ExportData(self, event):
        TemplateTextInputStart = "/* [wxMaxima: input   start ] */\n"
        TemplateTextInputEnd = "/* [wxMaxima: input   end    ] */\n\n\n"
        TemplateTextCommentStart = "/* [wxMaxima: comment start ]\n"
        TemplateTextCommentEnd = "   [wxMaxima: comment end    ] */\n\n\n"

        try:
            f = open(self.FSavePath.GetValue(), "w")
        except IOError:
            wx.LogError("Cannot save current data in file '%s'." %
 self.FSavePath.GetValue())

        f.write("/* [wxMaxima batch file version 1] [ DO NOT EDIT BY HAND!
 ]*/\n")
        f.write(TemplateTextInputStart)
        f.write('kill(all); \nload("../Library/my_funs.mac")$\n')
        f.write("fpprintprec : 6$ /* float numbers precision (only for
 printing) */\n")
        f.write("ratprint : false$\n")
        f.write(TemplateTextInputEnd)
        f.write(TemplateTextCommentStart)
        f.write("Definition of " + Problem.Type + " Elasticity Problem:\n")
        f.write('Nodal Coordinates "nodxy", Element Connectivity "elenod",
 Element Material Properties "elemat", ')
        if Problem.Method == "Numerical":
            f.write('Number of Quadrature Points "nqp", ')
```

```
    f.write('Number of Element Nodes "nen", Thickness Function "thickness", freedom tags list "nodtag", ')
    f.write('freedom values list "nodval", ')
    if Problem.Type == "Plane Stress" or Problem.Type == "Plane Strain":
        f.write('Edges defining the Neumann boundaries "edges", Pressures on the edges "pressures", ')
    f.write('Numerical or Analytical Calculations Switch "numerical"\n')
    f.write(TemplateTextCommentEnd)

    f.write(TemplateTextInputStart)

    NodeXY = "nodxy : ["
    for i in range(0, Tab2.NodeGrid.GetNumberRows()):
        NodeXY += "[" + Tab2.NodeGrid.GetCellValue(i, 0) + ", " + Tab2.NodeGrid.GetCellValue(i, 1) + "]"
        if i < Tab2.NodeGrid.GetNumberRows() - 1:
            NodeXY += ", "
    NodeXY += "]$\n"
    f.write(NodeXY)

    ElementNodes = "elenod : ["
    for i in range(0, Tab2.ElementGrid.GetNumberRows()):
        ElementNodes += "["
        for j in range(0, Problem.NEN):
            ElementNodes += Tab2.ElementGrid.GetCellValue(i, j + 1)
            if j < Problem.NEN - 1:
                ElementNodes += ", "
        ElementNodes += "]"
        if i < Tab2.ElementGrid.GetNumberRows() - 1:
            ElementNodes += ", "
    ElementNodes += "]$\n"
    f.write(ElementNodes)

    ElementMaterial = "elemat : ["
    try:
        for i in range(0, Tab2.ElementGrid.GetNumberRows()):
            ElementMaterial += "["
            ElementMatIndex = Tab2.ElementGrid.GetCellValue(i, 0)
            if Problem.Type == "Plane Stress" or Problem.Type == "Plane Strain" or Problem.Type == "Classical Plate":
                ElementMaterial += MaterialGrid.GetCellValue(0, int(ElementMatIndex) - 1) + ", " + \
                                   MaterialGrid.GetCellValue(1, int(ElementMatIndex) - 1) + "]"
            elif Problem.Type == "Shear Deformable Plate":
                ElementMaterial += MaterialGrid.GetCellValue(0, int(ElementMatIndex) - 1) + ", " + \
                                   MaterialGrid.GetCellValue(1, int(ElementMatIndex) - 1) + "," + \
                                   MaterialGrid.GetCellValue(2, int(ElementMatIndex) - 1) + "]"
            if i < Tab2.ElementGrid.GetNumberRows() - 1:
                ElementMaterial += ", "
    except:
        pass
    ElementMaterial += "]$\n"
    f.write(ElementMaterial)

    f.write("nummod : length(nodxy)$\n")
```

```
        f.write("numele : length(elenod)$\n\n")

        # Thickness
        f.write("thickness(x, y) := " + Problem.Thickness + "$\n\n")

        # Body forces
        if Problem.Type == "Plane Stress" or Problem.Type == "Plane Strain":
            f.write("body : [" + Problem.BodyX + ", " + Problem.BodyY +
"]$\n\n")
        elif Problem.Type == "Classical Plate":
            f.write("body_force(x, y) := " + Problem.Body + "$\n\n")
        elif Problem.Type == "Shear Deformable Plate":
            f.write("q_z(x, y) := " + Problem.QZ + "$\n")
            f.write("m_x(x, y) := " + Problem.MX + "$\n")
            f.write("m_y(x, y) := " + Problem.MY + "$\n\n")

            # Coefficients for the shear deformable plate
            f.write("/*\n")
            f.write("D_b : " + Problem.Material[0] + "*(" + Problem.Thickness
                    + ")^3/(12*(1-" + Problem.Material[1] + "^2))\n")
            f.write("D_s : (" + Problem.ShearCF + ")*(" + Problem.Material[2]
                    + ")*(" + Problem.Thickness + ")\n")
            f.write("*/\n\n")

        NodeTag = "nodtag : ["
        NodeVal = "nodval : ["
        for i in range(0, Tab3.NodeTagGrid.GetNumberRows()):
            NodeTag += "["
            NodeVal += "["
            for j in range(0, Problem.NDOF):
                if Tab3.NodeTagGrid.GetCellValue(i, j) != "":
                    NodeTag += Tab3.NodeTagGrid.GetCellValue(i, j)
                else:
                    NodeTag += str(0)

                if Tab3.NodeValGrid.GetCellValue(i, j) != "":
                    NodeVal += Tab3.NodeValGrid.GetCellValue(i, j)
                else:
                    NodeVal += str(0)

                if j < Problem.NDOF - 1:
                    NodeTag += ", "
                    NodeVal += ", "
            NodeTag += "]"
            NodeVal += "]"
            if i < Tab3.NodeTagGrid.GetNumberRows() - 1:
                NodeTag += ", "
                NodeVal += ", "
        NodeTag += "]$\n\n"
        NodeVal += "]$\n\n"

        f.write(NodeTag)
        f.write(NodeVal)

        if Problem.Type == "Plane Stress" or Problem.Type == "Plane Strain":
            edges = "edges : ["
            pressures = "pressures : ["
            for i in range(0, Problem.NEdge):
                edges += "[" + Tab2.EdgeGrid.GetCellValue(i, 0) + "]"
```

```
                pressures += "[" + Tab3.PressureGrid.GetCellValue(i, 0) + "]"
                if i < Problem.NEdge - 1:
                    edges += ", "
                    pressures += ", "
            edges += "]$\n\n"
            pressures += "]$\n\n"

            f.write(edges)
            f.write(pressures)

            # if Problem.Type == "Plane Stress" or Problem.Type == "Plane Strain":
            f.write('type : "' + Problem.Type.lower() + '"$\n')

        f.write("nen : " + str(Problem.NEN) + "$\n")

        if Problem.Method == "Numerical":
            f.write("numerical : true$\n")
            f.write("nqp : " + str(Problem.NQP) + "$\n")
        else:
            f.write("numerical : false$\n")

        f.write(TemplateTextInputEnd)

        f.write(TemplateTextCommentStart)
        f.write("Maxima Calculations and Output\n")
        f.write(TemplateTextCommentEnd)

        f.write(TemplateTextInputStart)
        f.write(Problem.get_solver())

        f.write('for i : 1 thru length(nodxy) do(\n')
        f.write('  print(" "),\n')
        f.write('  print("Displacements at Node ", i),\n')
        if Problem.Type == "Plane Stress" or Problem.Type == "Plane Strain":
            f.write('  print(" ", U_X, "=", float(noddis[i][1])),\n')
            f.write('  print(" ", U_Y, "=", float(noddis[i][2]))\n')
        elif Problem.Type == "Classical Plate" or Problem.Type == "Shear Deformable Plate":
            f.write('  print(" ", U_Z, "=", factor(ratsimp(noddis[i][1]))),\n')
            f.write('  print(" ", Phi_X, "=", factor(ratsimp(noddis[i][2]))),\n')
            f.write('  print(" ", Phi_Y, "=", factor(ratsimp(noddis[i][3])))\n')
        f.write(')$\n')

        f.write(TemplateTextInputEnd)

    # Event -Export Data - When the user click on "Export and open in wxMaxima"
    def ExportAndOpenData(self, event):
        self.ExportData(None)
        String = self.FOpenPath.GetValue() + ' -o ' + '"' + self.FSavePath.GetValue() + '"'
        try:
            os.popen(String)
        except:
            pass

# Problem information
```

```
class Problem():
    def __init__(self, type, method):
        # oinitialize all the required variables
        self.Type = type
        self.Method = method
        self.NDOF = 2
        self.Material = ["0", "0", "0"]
        self.NEN = 4
        self.Thickness = "t"
        self.NEdge = 0
        self.Body = "0"
        self.BodyX = "0"
        self.BodyY = "0"
        self.QZ = "0"
        self.MX = "0"
        self.MY = "0"
        self.NQP = "1"
        self.NUMNP = 1
        self.NUMEL = 1
        self.ShearCF = "0"
        self.NMat = 1
        self.NMatBox = None
        self.ThicknessBox = None
        self.ShearCFBox = None
        self.NUMNPBox = None
        self.NUMELBox = None
        self.NENBox = None
        self.NQPBox = None
        self.NEdgeBox = None
        self.BodyBox = None
        self.BodyXBox = None
        self.BodyYBox = None
        self.QZBox = None
        self.MXBox = None
        self.MYBox = None

    # Get the information about the solver line in wxMaxima
    def get_solver(self):
        Pre = "[noddis, Kmod, fmod] : "
        if self.Type == ('Plane Stress' or 'Plane Strain'):
            Vars = "(nodxy, elenod, elemat, edges, pressures, body, nodtag, nodval, numerical)$\n\n"
            SolverExpression = Pre + "PlanarElementsSolution" + Vars
        elif self.Type == "Classical Plate":
            Vars = "(nodxy, elenod, elemat, nodtag, nodval, numerical)$\n\n"
            SolverExpression = Pre + "KirschhoffPlateElementsSolution" + Vars
        elif self.Type == "Shear Deformable Plate":
            Vars = "(nodxy, elenod, elemat, nodtag, nodval, numerical)$\n\n"
            SolverExpression = Pre + "ReissnerMindlinPlateElementsSolution" + Vars
        else:
            SolverExpression = "problem type not supported"
        return SolverExpression

    # Get the information about the current problem (as a box)
    def GetProblemDescriptionBox(self):
        ProblemDescriptionBox = wx.BoxSizer(wx.VERTICAL)

        # Material Properties
```

```
   MaterialBox = wx.StaticBox(Tab1, -1, "Material Properties:")
   MaterialVerticalBox = wx.StaticBoxSizer(MaterialBox, wx.VERTICAL)
   mat_HorizontalBox = wx.BoxSizer(wx.HORIZONTAL)
   self.NMatBox = wx.TextCtrl(Tab1, -1)
   mat_HorizontalBox.AddMany([(wx.StaticText(Tab1, -1, "Number of
Material Sets:"), 0, wx.ALIGN_LEFT),
                     ((10, 10),), # this is a spacer
                     (self.NMatBox, 0, wx.ALIGN_LEFT |
wx.ALIGN_CENTER_VERTICAL)])
   MaterialVerticalBox.Add(mat_HorizontalBox)
   MaterialVerticalBox.AddSpacer(5)
   global MaterialGrid
   MaterialGrid = Tab1.CreateMaterialGrid()
   MaterialVerticalBox.Add(MaterialGrid)
   ProblemDescriptionBox.Add(MaterialVerticalBox)
   ProblemDescriptionBox.AddSpacer(10)

   # Geometrical Properties
   GeometryBox = wx.StaticBox(Tab1, -1, "Geometrical Properties:")
   GeometryBoxSizer = wx.StaticBoxSizer(GeometryBox, wx.VERTICAL)
   GeometryBoxFlexGrid = wx.FlexGridSizer(2, 3, 0, 0)
   self.ThicknessBox = wx.TextCtrl(Tab1, -1)
   GeometryBoxFlexGrid.AddMany([(wx.StaticText(Tab1, -1, "Thickness
(x,y):"), 0, wx.ALIGN_LEFT),
                     ((10, 10),), # this is a spacer
                     (self.ThicknessBox, 0, wx.ALIGN_LEFT |
wx.ALIGN_CENTER_VERTICAL)])
   if self.Type == "Shear Deformable Plate":
       self.ShearCFBox = wx.TextCtrl(Tab1, -1)
       GeometryBoxFlexGrid.AddMany([(wx.StaticText(Tab1, -1, "Shear
Correction Factor:"), 0, wx.ALIGN_LEFT),
                         ((10, 10),), # this is a spacer
                         (self.ShearCFBox, 0, wx.ALIGN_LEFT |
wx.ALIGN_CENTER_VERTICAL)])
       self.ShearCFBox.Bind(wx.EVT_TEXT, self.SetKs)
   GeometryBoxSizer.Add(GeometryBoxFlexGrid)
   ProblemDescriptionBox.Add(GeometryBoxSizer)
   ProblemDescriptionBox.AddSpacer(10)

   # Finite Element Model
   FEMBox = wx.StaticBox(Tab1, -1, "Finite Element Model:")
   FEMBoxSizer = wx.StaticBoxSizer(FEMBox, wx.VERTICAL)
   FEMBoxFlexGrid = wx.FlexGridSizer(5, 3, 0, 0)
   self.NUMNPBox = wx.TextCtrl(Tab1, -1)
   self.NUMELBox = wx.TextCtrl(Tab1, -1)
   self.NENBox = wx.TextCtrl(Tab1, -1)
   FEMBoxFlexGrid.AddMany([(wx.StaticText(Tab1, -1, "Total Number of
Nodes (numnp):"), 0, wx.ALIGN_LEFT),
                    ((10, 10),), # this is a spacer
                    (self.NUMNPBox, 0, wx.ALIGN_LEFT |
wx.ALIGN_CENTER_VERTICAL),
                    (wx.StaticText(Tab1, -1, "Total Number of Elements
(numel):"), 0, wx.ALIGN_LEFT),
                    ((10, 10),), # this is a spacer
                    (self.NUMELBox, 0, wx.ALIGN_LEFT |
wx.ALIGN_CENTER_VERTICAL),
                    (wx.StaticText(Tab1, -1, "Number of Element Nodes
(nen):"), 0, wx.ALIGN_LEFT),
                    ((10, 10),), # this is a spacer
```

```
                        (self.NENBox, 0, wx.ALIGN_LEFT |
 wx.ALIGN_CENTER_VERTICAL)
                        ])
    if self.Method == "Numerical":
        self.NQPBox = wx.TextCtrl(Tab1, -1)
        FEMBoxFlexGrid.AddMany([(wx.StaticText(Tab1, -1, "Number of
 Quadrature Points (nqp):"), 0, wx.ALIGN_LEFT),
                             ((10, 10),), # this is a spacer
                             (self.NQPBox, 0, wx.ALIGN_LEFT |
 wx.ALIGN_CENTER_VERTICAL)])
        # Event Handler
        self.NQPBox.Bind(wx.EVT_TEXT, self.SetNqp)
    if self.Type == "Plane Stress" or self.Type == "Plane Strain":
        Tab2.Sizer.Show(Tab2.Sizer.GetItemCount() - 1)
        Tab3.Sizer.Show(Tab3.Sizer.GetItemCount() - 1)
        self.NEdgeBox = wx.TextCtrl(Tab1, -1)
        FEMBoxFlexGrid.AddMany([(wx.StaticText(Tab1, -1, "Number of
 Neumann Edges (nedge):"), 0, wx.ALIGN_LEFT),
                             ((10, 10),), # this is a spacer
                             (self.NEdgeBox, 0, wx.ALIGN_LEFT |
 wx.ALIGN_CENTER_VERTICAL)])
        # Event Handlers
        self.NEdgeBox.Bind(wx.EVT_TEXT, Tab2.SetMeshGridNedge)
        self.NEdgeBox.Bind(wx.EVT_TEXT, Tab3.SetPressureGridNedge)
        self.NEdgeBox.Bind(wx.EVT_TEXT, self.SetNedge)
    else:
        Tab2.Sizer.Hide(Tab2.Sizer.GetItemCount() - 1)
        Tab3.Sizer.Hide(Tab3.Sizer.GetItemCount() - 1)

    FEMBoxSizer.Add(FEMBoxFlexGrid)
    ProblemDescriptionBox.Add(FEMBoxSizer)
    ProblemDescriptionBox.AddSpacer(10)

    # Body Forces
    BodyBox = wx.StaticBox(Tab1, -1, "Body Forces:")
    BodyBoxSizer = wx.StaticBoxSizer(BodyBox, wx.VERTICAL)
    BodyBoxFlexGrid = wx.FlexGridSizer(4, 3, 0, 0)
    if self.Type == "Plane Stress" or self.Type == "Plane Strain":
        self.BodyXBox = wx.TextCtrl(Tab1, -1)
        self.BodyYBox = wx.TextCtrl(Tab1, -1)
        BodyBoxFlexGrid.AddMany([(wx.StaticText(Tab1, -1, "Body Force
 (x):"), 0, wx.ALIGN_LEFT),
                             ((10, 10),), # this is a spacer
                             (self.BodyXBox, 0, wx.ALIGN_LEFT |
 wx.ALIGN_CENTER_VERTICAL),
                             (wx.StaticText(Tab1, -1, "Body Force (y):"), 0,
 wx.ALIGN_LEFT),
                             ((10, 10),), # this is a spacer
                             (self.BodyYBox, 0, wx.ALIGN_LEFT |
 wx.ALIGN_CENTER_VERTICAL)
                             ])
        # Event Handlers
        self.BodyXBox.Bind(wx.EVT_TEXT, self.SetBodyX)
        self.BodyYBox.Bind(wx.EVT_TEXT, self.SetBodyY)
    elif self.Type == "Classical Plate":
        self.BodyBox = wx.TextCtrl(Tab1, -1)
        BodyBoxFlexGrid.AddMany([(wx.StaticText(Tab1, -1, "Body Force
 (x,y):"), 0, wx.ALIGN_LEFT),
```

```
                                  (self.BodyBox, 0, wx.ALIGN_LEFT | wx.ALIGN_CENTER_VERTICAL)
                                  ])
            # Event Handler
            self.BodyBox.Bind(wx.EVT_TEXT, self.SetBody)
        elif self.Type == "Shear Deformable Plate":
            self.QZBox = wx.TextCtrl(Tab1, -1)
            self.MXBox = wx.TextCtrl(Tab1, -1)
            self.MYBox = wx.TextCtrl(Tab1, -1)
            BodyBoxFlexGrid.AddMany([(wx.StaticText(Tab1, -1, "q_z (x,y):"), 0, wx.ALIGN_LEFT),
                                  ((10, 10),), # this is a spacer
                                  (self.QZBox, 0, wx.ALIGN_LEFT | wx.ALIGN_CENTER_VERTICAL),
                                  (wx.StaticText(Tab1, -1, "m_x (x,y):"), 0, wx.ALIGN_LEFT),
                                  ((10, 10),), # this is a spacer
                                  (self.MXBox, 0, wx.ALIGN_LEFT | wx.ALIGN_CENTER_VERTICAL),
                                  (wx.StaticText(Tab1, -1, "m_y (x,y):"), 0, wx.ALIGN_LEFT),
                                  ((10, 10),), # this is a spacer
                                  (self.MYBox, 0, wx.ALIGN_LEFT | wx.ALIGN_CENTER_VERTICAL)
                                  ])
            # Event Handlers
            self.QZBox.Bind(wx.EVT_TEXT, self.SetQz)
            self.MXBox.Bind(wx.EVT_TEXT, self.SetMx)
            self.MYBox.Bind(wx.EVT_TEXT, self.SetMy)

        BodyBoxSizer.Add(BodyBoxFlexGrid)
        ProblemDescriptionBox.Add(BodyBoxSizer)

        # Event Handlers
        self.NUMNPBox.Bind(wx.EVT_TEXT, Tab2.SetNodeGridNumnp)
        self.NUMNPBox.Bind(wx.EVT_TEXT, Tab3.SetNodeTagValGridNdof)
        self.NUMNPBox.Bind(wx.EVT_TEXT, self.SetNumnp)
        self.NUMELBox.Bind(wx.EVT_TEXT, Tab2.SetElementGridNumel)
        self.NUMELBox.Bind(wx.EVT_TEXT, self.SetNumel)
        self.NENBox.Bind(wx.EVT_TEXT, Tab2.SetElementGridNen)
        self.NENBox.Bind(wx.EVT_TEXT, self.SetNen)
        Tab1.Bind(wx.EVT_TEXT, lambda event: Tab1.UpdateMaterialGrid(event, MaterialGrid), self.NMatBox)
        self.NMatBox.Bind(wx.EVT_TEXT, self.SetNmat)
        self.ThicknessBox.Bind(wx.EVT_TEXT, self.SetThickness)

        return ProblemDescriptionBox

    # Event - Update the value of body forces
    def SetBody(self, event):
        Object = event.GetEventObject()
        self.Body = Object.GetValue()

    # Event - Update the value of body forces - x-direction
    def SetBodyX(self, event):
        Object = event.GetEventObject()
        self.BodyX = Object.GetValue()

    # Event - Update the value of body forces - y-direction
```

```
    def SetBodyY(self, event):
        Object = event.GetEventObject()
        self.BodyY = Object.GetValue()

    # Event - Update the value of Ks
    def SetKs(self, event):
        Object = event.GetEventObject()
        self.ShearCF = Object.GetValue()

    # Event - Update the value of the Poisson's ratio
    def SetMaterialPropertiesPoisson(self, event):
        Object = event.GetEventObject()
        self.Material[1] = Object.GetValue()

    # Event - Update the value of the shear modulus
    def SetMaterialPropertiesShearModulus(self, event):
        Object = event.GetEventObject()
        self.Material[2] = Object.GetValue()

    # Event - Update the value of the Young's modulus
    def SetMaterialPropertiesYoung(self, event):
        Object = event.GetEventObject()
        self.Material[0] = Object.GetValue()

    # Event - Update the value of Mx
    def SetMx(self, event):
        Object = event.GetEventObject()
        self.MX = Object.GetValue()

    # Event - Update the value of My
    def SetMy(self, event):
        Object = event.GetEventObject()
        self.MY = Object.GetValue()

    # Event - Update the value of NDOF
    def SetNdof(self, event):
        Object = event.GetEventObject()
        problem = Object.GetStringSelection()
        if problem == "Plane Stress" or problem == "Plane Strain":
            self.NDOF = 2
        else:
            self.NDOF = 3
        event.Skip()

    # Update the value of NDOF (not an event)
    def SetNdofNoEvent(self):
        if self.Type == "Plane Stress" or self.Type == "Plane Strain":
            self.NDOF = 2
        else:
            self.NDOF = 3

    # Event - Update the value of NEdge
    def SetNedge(self, event):
        Object = event.GetEventObject()
        self.NEdge = int(Object.GetValue())
        event.Skip()

    # Event - Update the value of NEN
    def SetNen(self, event):
```

```
        Object = event.GetEventObject()
        self.NEN = int(Object.GetValue())
        event.Skip()

    # Event - Update the value of NMat
    def SetNmat(self, event):
        Object = event.GetEventObject()
        if Object.GetValue() != '':
            self.NMat = int(Object.GetValue())
        event.Skip()

    # Event - Update the value of NQP
    def SetNqp(self, event):
        Object = event.GetEventObject()
        self.NQP = int(Object.GetValue())

    # Event - Update the value of NUMEL
    def SetNumel(self, event):
        Object = event.GetEventObject()
        self.NUMEL = int(Object.GetValue())
        event.Skip()

    # Event - Update the value of NUMNP
    def SetNumnp(self, event):
        Object = event.GetEventObject()
        self.NUMNP = int(Object.GetValue())
        event.Skip()

    # Event - Update the value of Qz
    def SetQz(self, event):
        Object = event.GetEventObject()
        self.QZ = Object.GetValue()

    # Event - Update the value of thickness
    def SetThickness(self, event):
        Object = event.GetEventObject()
        self.Thickness = Object.GetValue()


# Main Frame of the application
class MainFrame(wx.Frame):
    def __init__(self):
        wx.Frame.__init__(self, None, title="wxMaxima Input Generator",
     size=wx.Size(1200, 750))
        self.SetMinSize((600, 150))

        global Problem, Tab1, Tab2, Tab3, Tab4

        # Initialize the problem
        Problem = Problem("None", "Analytical")

        # Create a panel and notebook (tabs holder)
        panel = wx.Panel(self, -1, style=wx.SIMPLE_BORDER)
        notebook = wx.Notebook(panel)

        # List of available problem types and solution methods
        problemList = ['None', 'Plane Stress', 'Plane Strain', 'Classical
     Plate', 'Shear Deformable Plate']
        methodsList = ['Analytical', 'Numerical']
```

```

        # Create the tab windows
        Tab1 = TabOne(notebook, problemList, methodsList)
        Tab2 = TabTwo(notebook)
        Tab3 = TabThree(notebook)
        Tab4 = TabFour(notebook)

        # Add the windows to tabs and name them
        notebook.AddPage(Tab1, "Problem Description")
        notebook.AddPage(Tab2, "Nodal Coordinates and Element Connectivity")
        notebook.AddPage(Tab3, "Boundary Conditions")
        notebook.AddPage(Tab4, "Output")

        # Set notebook in a sizer to create the layout
        sizer = wx.BoxSizer(wx.VERTICAL)
        sizer.Add(notebook, 1, wx.EXPAND)
        panel.SetSize((1000, 500))
        panel.SetSizer(sizer)

        # Status bar
        self.statusbar = self.CreateStatusBar(2)
        self.statusbar.SetStatusText('Select a problem type', 0)
        self.statusbar.SetStatusText('Select a solution method', 1)

        # Event handlers
        RadioBox1.Bind(wx.EVT_RADIOBOX, self.SetVal)
        RadioBox2.Bind(wx.EVT_RADIOBOX, self.SetVal)
        RadioBox1.Bind(wx.EVT_RADIOBOX, Problem.SetNdof)

    # Set the status bar text
    def SetStatusBar(self):
        if RadioBox1.GetStringSelection() == "None":
            self.statusbar.SetStatusText('Select a problem type', 0)
            Tab1.Sizer.Hide(2)
            Tab1.Layout()
        else:
            self.statusbar.SetStatusText(Problem.Type + " Problem", 0)
            self.statusbar.SetStatusText(Problem.Method + " Solution", 1)

    # Event - Set the problem type and solution method
    def SetVal(self, event):
        Object = event.GetEventObject()
        if Object.GetStringSelection() == "None":

            self.statusbar.SetStatusText('Select a problem type', 0)

            Tab1.Sizer.Hide(2)
            Tab1.Layout()
        else:
            Problem.Type = RadioBox1.GetStringSelection()
            Problem.Method = RadioBox2.GetStringSelection()
            self.SetStatusBar()

            Tab1.ShowProperties()

            if Problem.Type == "Plane Stress" or Problem.Type == "Plane
     Strain":
                while Tab2.EdgeBoxSizer.GetItemCount() > 0:
     Tab2.EdgeBoxSizer.Hide(Tab2.EdgeBoxSizer.GetItemCount() - 1)
```

```
                    Tab2.EdgeBoxSizer.Detach(Tab2.EdgeBoxSizer.GetItemCount() - 1)

                while Tab3.PressureBoxSizer.GetItemCount() > 0:
                    Tab3.PressureBoxSizer.Hide(Tab3.PressureBoxSizer.GetItemCount() - 1)
                    Tab3.PressureBoxSizer.Detach(Tab3.PressureBoxSizer.GetItemCount() - 1)

                Tab2.AddMeshEdges(event)
                Tab3.AddPressureEdges(event)

        event.Skip()

if __name__ == "__main__":
    App = wx.App()
    MainFrame = MainFrame()
    MainFrame.Show()
    App.MainLoop()
```

Reference

1. wxPython - The GUI toolkit for Python. https://wxpython.org. Accessed 20 Mar 2021

Index

A. Öchsner and R. Makvandi, *Plane Finite Elements for Two-Dimensional Problems*,
https://doi.org/10.1007/978-3-030-89550-1